Contributions of
African American
Scientists and Mathematicians

Principal Researcher

Mozell P. Lang

Former Science Consultant
Michigan Department of Education
Science Consultant, Highland Park Schools
Highland Park, Michigan

Advisers and Reviewers

Thelma Gardner

Former Elementary Science Supervisor
Detroit Public Schools
Adjunct Faculty, Wayne State University
Detroit, Michigan

Napoleon Adebola Bryant, Jr.

Professor Emeritus of Education
Xavier University
Cincinnati, Ohio

Harcourt
SCHOOL PUBLISHERS

Orlando   Boston   Dallas   Chicago   San Diego
www.harcourtschool.com

Photo Credits

Kindergarten – Grade 1

pp1, 3, 5, 7, 9, 11, 13, 15, Illustrations by Donovan Foote; p11, Moorland-Spingarn/Howard University; p15, The Gazette

Grade 2

p17, Library of Congress; p19, Illustration by Donovan Foote; p21, Photos Courtesy of the Jemison Group/ NASA; p23, Library of Congress, Image by Anne McCarthy-Courtesy the Lemelson-MIT program; p27, Texas Southern University; p29, Courtesy of Diane Fenster of SFSU; p31, Getty Images, Library of Congress

Grade 3

pp33, 37, 45, 47, Illustrations by Donovan Foote; p35, Photos Courtesy of Dr. Canady; p39, Photos Courtesy of www.emeagwali.com; p41, NASA Langley Research Center; p43, RPI/Mark McCarty, NASA; p49, Corbis; p53, Photos Courtesy of James Tilmon; p55, Photos Courtesy of Copyright University Corporation for Atmospheric Research

Grade 4

p57, Photos Courtesy of Lenore Blum; p59, United States Navy, NASA; pp61, 75, 87, Illustrations by Donovan Foote; p63, With permission from Dr. Farley; p65, National Oceanic and Atmospheric Administration/Atlantic Oceanographic and Meteorological Laboratory, Miami; p69, Photos Courtesy of the Thermo-King Corporation; p71, Moorland-Spingarn/Howard University; p73, NASA; p75, Library of Congress; p77, Photos Courtesy of Katherine Okikiolu; p79, U.S. Geological Survey; p81, Photos Courtesy of Dr. Petters, NASA; p83, Photos Courtesy of Dr. Ross-Lee; p85, Photos Courtesy of Dr. Williams, NASA

Grade 5

p89, Maryland Historical Society; pp91, 93, 97, 107, 109, 113, 115, 119 Illustrations by Donovan Foote; p95, California State University at Fullerton, Corbis; p99, Photos Courtesy of Christine Darden/NASA; p101, Image reproduced by permission of IBM Research, Almaden Research Center, unauthorized use not permitted; p103, Photos Courtesy of Nathaniel Dean; p105, National Portrait Gallery, Smithsonian Institution, Library of Congress; p111, National Portrait Gallery, Smithsonian Institution; p117, Moorland-Spingarn/Howard University

Grade 6

p121, State Historical Society of Iowa, Courtesy National Archives; p123, NASA with permission from Guion Bluford; p125, Photos Courtesy of Dr. Ben Carson; p127, Photo Courtesy of www.emeagwali.com; p129, Image by Hillary Mitchell-Courtesy the Lemelson-MIT program; p131, Library of Congress; p133, Photo Courtesy of Lenore Blum; p137, Photos Courtesy of Lenore Blum; p139, Photos from the Archives and Special Collections of DePauw University and Indiana United Methodist; pp141, 145, Illustrations by Donovan Foote; p143, Photo Courtesy of Dr. Okpodu; p147, Photos Courtesy of Charles Abramson, Photo Source USDA; p149, Dennis Finnin, AMNH, Delvinhair Productions; p151, Photo Courtesy of Lenore Blum

Contents

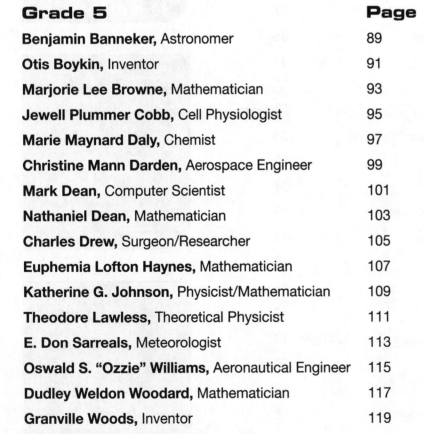

Grade 6

To the Teacher

Contributions of African American Scientists and Mathematicians, a resource book for grades K–6, is designed to assist you in providing positive role models for students, while at the same time linking the work of actual scientists and mathematicians to specific concepts you are teaching from science and mathematics programs by Harcourt School Publishers. Each of the 76 biographies provides students with the opportunity to do research on the individual discussed or to extend the concepts presented.

Lesson Content

For the Students in Grades K, 1, or 2

Lessons for K–2 students include an activity page and a teacher resource page. The teacher resource page includes prompts to assess and extend learning as well as a section called *Reading Aloud* to introduce the scientist or mathematician and their field of study.

For the Students in Grades 3, 4, 5, or 6

Each lesson consists of the following: an explanation of the scientist's or mathematician's work and personal background; *Explain It!,* discussion questions that focus on contributions made by the individual; and *Explore It!,* an idea for extension of the lesson focusing on a project, research ideas or careers.

For the Teacher

The lesson plan for the teacher includes the following elements:

- *Purpose*
- *Science Background*
- *Introduce,* an idea for introducing the lesson
- *Discuss,* discussion questions and ideas
- *Extend,* which may include ideas for project research
- *Career Focus,* a question for students that relates to a career in science

Professional Bibliography

This bibliography lists some of the sources that were used to create **Contributions of African American Scientists and Mathematicians.** It includes professional resources and children's resources for multicultural education. This bibliography is not meant to serve as an all-inclusive list of resources.

Following the bibliography is a list of additional organizations the editors used. These institutions were particularly helpful in procuring information on the careers and lives of the African Americans featured here.

Adair, Gene, and Nathan Irvin Huggins. *George Washington Carver: Botanist.* Philadelphia, PA: Chelsea House Publications, 1989.

Beckner, Chrisanne. *100 African Americans Who Shaped American History.* San Mateo, CA: Bluewood Books, 1995.

Bernstein, Leonard, Alan Winkler, and Linda Zierdt–Warshaw. *African and African American Women of Science.* Saddle Brook, NJ: Peoples Publishing Group, Inc., 1997.

Brodie, James Michael. *Created Equal: The Lives and Ideas of Black American Innovators.* New York, NY: William Morrow & Co., 1993.

Cerami, Charles, and Robert M. Silverstein. *Benjamin Banneker: Surveyor, Astronomer, Publisher, Patriot.* Indianapolis, IN: Wiley, 2002.

Davis, Marianna W., ed. *Contributions of Black Women to America.* Volume II: Civil Rights, Politics and Government, Education, Medicine, Sciences. Columbia, SC: Kenday Press, Inc., 1982.

Donovan, Richard X. *African-American Scientists.* Portland, OR: National Book Company, 1999.

Driver, Paul J. *Black Giants in Science.* New York, NY: Vantage Press, 1978.

Fouche, Rayvon. *Black Inventors in the Age of Segregation: Granville T. Woods, Lewis H. Latimer, and Shelby J. Davidson.* Baltimore, MD: The Johns Hopkins University Press, 2003.

Gubert, Betty Kaplan, Miriam Sawyer, and Caroline M. Fannin. *Distinguished African Americans in Aviation and Space Science.* Westport, CT: Oryx Press, 2001.

Haber, Louis. *Black Pioneers of Science and Invention.* New York, NY: Odyssey Classics, 1992.

Harley, Sharon. *Timetables of African-American History: A Chronology of the Most Important People and Events in African-American History.* New York, NY: Touchstone Publishers, 1996.

Haskins, Jim. *Outward Dreams: Black Inventors and Their Inventions.* New York, NY: Walker & Co., 2003.

Hudson, Wade. *Book of Black Heroes: Scientists, Healers, and Inventors.* East Orange, NJ: Just Us Books, 2003.

Jemison, Mae. *Find Where the Wind Goes: Moments From My Life.* New York, NY: Scholastic, Inc., 2002.

Jenkins, Edward S., Patricia Stohr–Hunt, Exyie C. Ryder, and S. Maxwell Hines. *To Fathom More: African American Scientists and Inventors.* Lanham, MD: University Press of America, 1996.

Jones, Stanley P., Jetty Kahn, and Fred M. Amram. *African-American Inventors.* Mankato, MN: Capstone Press, 1996.

Kessler, James H., Jerry S. Kidd, and Renee A. Kidd. *Distinguished African American Scientists of the 20th Century.* Westport, CT: Greenwood Publishing Group, Inc., 1996.

Krapp, Kristine M., *Notable Black American Scientists.* Stamford, CT: Thomson Gale, 1998.

March, Wina. *African American Achievers in Science, Medicine, and Technology: A Resource Book for Young Learners, Parents, Teachers, and Librarians.* Bloomington, IN: Authorhouse, 2003.

McClure, Judy. *Remarkable Women: Past and Present, Healer and Researchers.* Austin, TX: Raintree Steck–Vaughn, 2000.

Miles, Johnnie H., et al. *Almanac of African American Heritage.* Indianapolis, IN: Jossey-Bass, 2001.

Rennert, Richard Scott, and Coretta Scott King. *Pioneers of Discovery.* Philadelphia, PA: Chelsea House Publications, 1994.

Sammons, Vivian, ed. *Blacks in Science and Medicine.* New York, NY: Hemisphere Publishing Co., 1989.

Schraff, Anne E. *Dr. Charles Drew: Blood Bank Innovator.* Berkeley Heights, NJ: Enslow Publishers, Inc., 2003.

Sluby, Patricia Carter. *The Inventive Spirit of African Americans: Patented Ingenuity.* Westport, CT: Praeger Publishers, 2004.

Spangenburg, Ray, and Kit Moser. *African Americans in Science, Math, and Invention.* New York, NY: Facts on File, Inc., 2003.

St. John, Jetty, Susan K. Henderson, and Fred M. Amram. *African-American Scientists.* Mankato, MN: Capstone Press, 1996.

Stille, Darlene R. *Extraordinary Women Scientists.* Chicago, IL: Childrens Press, 1995.

Sullivan, Otha Richard, and Jim Haskins. *Black Stars: African American Inventors.* Indianapolis, IN: Wiley, 1998.

Sullivan, Otha Richard, and Jim Haskins. *Black Stars: African American Women Scientists and Inventors.* Indianapolis, IN: Wiley, 2001.

Tyson, Neil deGrasse. *The Sky Is Not the Limit: Adventures of an Urban Astrophysicist.* Amherst, NY: Prometheus Books, 2004.

Watson, Clifford. *Ten Great African American Men of Science: With Hands-On Activities.* Saddle Brook, NJ: Peoples Publishing Group, Inc., 1995.

Webster, Raymond B. *African American Firsts in Science & Technology.* Detroit, MI: Gale Group, 1999.

Wilson, Donald, and Jane Wilson. *The Pride of African American History: Inventors, Scientists, Physicians, Engineers: Featuring Many Outstanding African Americans.* Bloomington, IN: Authorhouse, 2003.

Yount, Lisa. *Black Scientists (American Profiles).* New York, NY: Facts on File, Inc., 1991.

Additional Sources

Association for Women in Science, *www.awis.org*

Auburn Avenue Research Library, *www.af.public.lib.ga.us/aarl/*

Carnegie Library of Pittsburgh, *www.clpgh.org/*

Moorland-Spingarn Research Center, *www.founders.howard.edu/moorland-spingarn/default.htm*

National Archives for Black Women's History, *www.nps.gov/mamc/bethune/archives/main.htm*

Schomburg Center for Research in Black Culture, *www.nypl.org/research/sc/sc.html*

W.E.B. Du Bois Institute for African and African American Research, *www.fas.harvard.edu/~du_bois/*

Andrew **Beard**

How do train cars stay together?
Train cars are joined by a hook called the Jenny
Coupler. Draw a circle around it.
Andrew Beard made the Jenny Coupler.
Color the picture.

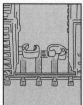

Andrew **Beard**
TEACHING NOTES

Purpose: To introduce students to the inventor of the Jenny Coupler.

Science Background: The dangerous job of hooking railroad cars together once involved a worker manually dropping a pin at the precise moment that two cars bumped into each other. The connection could not be off by even a second or a worker could become injured or even crushed to death between the cars. The Jenny Coupler was a new device that could do this more easily and safely than the manual method.

I. INTRODUCE

About Andrew Beard: Andrew Beard was born a slave. He had many jobs growing up. As a farmer in Alabama, Beard built and patented his first invention, a plow.

In 1892, while working on the railroad, his leg became smashed between two train cars and he had to have his leg amputated. While getting used to his amputated leg, Beard got the idea of how to prevent people from injury while working on the train.

About Beard's Work: Andrew Beard invented machines, such as the Jenny Coupler, that made work easier and saved time and lives. Beard also invented a rotary steam engine in 1899.

2. DISCUSS

Use the copy master to prompt discussion. Lead students to understand that inventors are problem solvers. They use the laws of math and science to figure out practical solutions to problems. Their job is to think up ways to use power and materials to make the world a better, safer, and more efficient place. The Jenny Coupler made the job of a railroad worker safer.

3. EXTEND

Discuss other inventions that have made the world a better and safer place.

reading aloud

Bump! went the train.

"What was that?" asked Jeffrey. He was standing outside the train waiting to board with his mother.

"Don't worry," said the engineer. "That noise is the sound of two train cars joining together. When the cars bump into each other, they lock together so they will stay together during the train ride."

"What holds them together?" Jeffrey asked.

"These do," said the engineer as he pointed to two hooks that looked like two people holding hands.

Andrew Beard invented these "hooks" in the 1890s. His invention changed the railroad industry. It also saved lives and prevented many injuries. That's because hooking train cars together by hand was a dangerous job. Today, engineers use automatic couplers.

June **Bacon-Bercey**

**June Bacon-Bercey is a meteorologist.
She helps people learn about weather.
Helping people learn about the weather is important.
It can help them stay safe.
Color the clouds in the sky.
Make one cloud a storm cloud.**

June **Bacon-Bercey**
TEACHING NOTES

Purpose: To introduce students to a meteorologist who works to accurately predict and help people understand the weather.

Science Background: In order to tell others about the weather, meteorologists collect information from around the world and use it to predict and explain how the atmosphere affects Earth and life on the planet.

Meteorologists study information on air pressure, temperature, humidity, and wind velocity to make their predictions. They also use clues in the environment to help them with their understanding.

Clouds offer clues as to what kind of weather is coming. Clouds will change their color, shape, and size depending upon the changes in the atmosphere and what type of weather is coming.

I. INTRODUCE

About June Bacon-Bercey: June Bacon-Bercey, a native of Kansas, always wanted to be a meteorologist—someone who studies the weather. In school, she studied math and meteorology.

About Bacon-Bercey's Work: Bacon-Bercey started her career as a TV weather forecaster before becoming chief administrator of TV activities for the National Oceanic and Atmospheric Administration (NOAA). She organized the services that NOAA provides to schools, colleges, professional and technical organizations, the government, and the general public that help people explore, understand, observe, or forecast the weather accurately. She then joined the National Weather Service, where she worked with air traffic controllers. Bacon-Bercey was presented with an award from the National Science Foundation for her work in the field of meteorology.

2. DISCUSS

Use the copy master to prompt discussion. Discuss weather with students. Help them understand how people depend on the weather forecast to stay safe and do their jobs. Explain that people make decisions about when to plant crops, when to stay out of the water, and what type of clothing to wear. What decisions do students make based on a weather report?

3. EXTEND

Have students draw pictures of the weather for each day of the week. Post them on a weather bulletin board, and discuss their observations.

reading aloud

Ailene's class is going on a picnic today. When Ailene got up this morning, she looked out the window and saw that it had rained. There were lots of gray clouds in the sky. Her mom told her that they should watch the weather report on television to find out if it would be a sunny or rainy day. Ailene and her mother turned on the television and listened to the meteorologist give the weather report. It was going to clear up and be sunny! Hooray! They would have a picnic after all.

Many years ago when people wanted to know what the weather would be like, they looked for clues. One clue was the kind of clouds that they saw in the sky. Today, meteorologists like June Bacon-Bercey use computers and other equipment to tell them about the weather. Bacon-Bercey enjoys her job because she helps people by giving them the information that they need to do their jobs or to stay safe. Because she likes her job so much, June Bacon-Bercey encourages young women and minorities to study science in school.

Elizabeth "Bessie" Coleman

Bessie Coleman was a pilot.
She flew planes.
She liked to do tricks in her planes.
Color the picture. Circle the pilot.

Elizabeth "Bessie"
Coleman
TEACHING NOTES

Purpose: To introduce students to an aviator who was the first African-American woman to receive her pilot's license.

Science Background: Elizabeth "Bessie" Coleman was born in Texas in 1892. Air shows were popular forms of entertainment when Coleman was a young woman. They moved from place to place and featured pilots known as *barnstormers*. Barnstormers learned how to fly aerial acrobatics, or stunts.

Coleman performed her stunts in World War I planes called "Jenny" trainers or "Jennies" (U.S. Army Curtiss JN-4) and DeHavilands.

1. INTRODUCE

About Bessie Coleman: When Bessie Coleman was a young woman, she dreamed about learning how to fly. She applied to different flying schools but none of them would allow her to attend because she was an African-American woman.

With the help of two people—Robert A. Abbott and Jesse Binga—Coleman was able to go to flying school in France.

Coleman was killed in an airplane crash in Florida in 1926.

About Coleman's Work: Bessie Coleman became the first woman to earn an international aviation license and the world's first African-American woman to receive her pilot's license. When Bessie returned from flight school in France, she began working as a performer at air shows. While she traveled from place to place, she spoke at churches, schools, and theaters about encouraging African Americans, especially women, to fly. She worked to start a flight school where other women and African Americans could learn to fly.

In 1995, the U.S. Postal Service honored Coleman by issuing a stamp commemorating a true "American legend."

2. DISCUSS

Use the copy master to prompt discussion. Lead students to discuss how they think pilots learn how to perform stunts.

3. EXTEND

Ask a librarian to provide students with books about airplanes, jets, the space shuttle, and any other machines that pilots fly. Have students make a picture of their favorite flying machine.

◄◄ reading aloud

Cheryl is very excited! Today she is going to the state fair to watch the air show. Cheryl likes the air show the best. The pilots do tricks. Someday Cheryl wants to learn how to fly a plane.

Bessie Coleman was a pilot who performed stunts while flying planes. In fact, she was the first African-American woman to receive a pilot's license. When her brother returned from fighting in World War I, he told her all about the new planes he saw. Bessie knew she had to learn how to fly. She read everything she could about planes. Finally, she was able to take flying lessons in France.

When she returned from France, she was famous. Since there were few jobs for pilots at that time, and even fewer for women and African-American pilots, Coleman began working as a performer at air shows. She used her fame to fight for equal rights for African Americans. She often would not fly in the show in places where African Americans were not allowed.

Clarence "Skip" Ellis

The picture shows Dr. Ellis with a computer screen.
Draw a circle around the computer's keyboard.
Color the picture.

Clarence "Skip" Ellis
TEACHING NOTES

Purpose: To introduce students to a mathematician whose specialty is computer science.

Science Background: Punch cards were first used to get data into computers. The cards had rows of numerals, letters, and sometimes symbols. Each number, letter, or symbol that was punched when it was put into the computer represented data to be stored and remembered. Over time, data was stored on tape and then on computer discs. Today, computers are able to store the data that people type on the keyboard. People can also make copies of their stored files on CDs and high-capacity drives.

1. INTRODUCE

About Clarence "Skip" Ellis: Dr. Clarence Ellis was born in Chicago in 1943. Dr. Ellis first became interested in computers when he took a job at a local company at age 15. He was so interested in computers that he studied computer science at Beloit College in Wisconsin. While he was attending Beloit, a computer was donated to the school. Dr. Ellis and his chemistry professor set up the school's first computer. This was

the beginning of the school's computer lab. Ellis attended graduate school at the University of Illinois, where he became the first African American to receive a Ph.D. in computer science.

About Ellis' Work: Dr. Ellis is now a professor of computer science and director of a research group at the University of Colorado. His work has helped develop basic software used by every computer in the world. Ask students what they think Dr. Ellis enjoyed learning in school.

2. DISCUSS

Use the copy master to prompt discussion. To help students understand how punch cards stored data, make a card for each student that has the letters of the alphabet and the numbers 1 to 10 on it. Help students use a hole punch to punch out the letters that spell his/her first name, and the number representing his/her age. Explain that punch cards stored data like this. How are the cards the same as the students'? How are they different?

3. EXTEND

Have students share what they use computers for.

reading aloud

Dr. "Skip" Ellis was one of five children. His family was very poor. When he was 15, he got a job at a local company. The money he earned helped to take care of his family.

Dr. Ellis was a night watchman. He had to make sure that no one broke in. He also watched over the company's new computer. In 1958 computers were very expensive and not many people had one. Dr. Ellis read the manual that came with the new computer. This was how he first learned so much about computers. One day, there was a problem with the computer. They had run out of punch cards. Early computers used punch cards to enter data. Without new punch cards, the computer was useless. Dr. Ellis was the only one who knew how to reuse old cards. He changed some settings on the computer, and the old cards worked great.

Dr. Ellis has written many articles about computers. He is a teacher and helps people work with computers.

Mary Styles **Harris**

Mary Styles Harris is a scientist. She studies cells in the human body.
Dr. Harris looks for clues on why people look and act like they do.
Circle the tool that she is using.

Mary Styles Harris
TEACHING NOTES

Purpose: To introduce students to a geneticist who works to help people understand the human body and the science of heredity.

Science Background: Genetics is the study of heredity, or how traits of organisms are passed down to their offspring. Genetics also determines how traits are lost, changed, or mutate over time.

Genetics focuses primarily on genes—codes found in the chromosomes. Chromosomes are the "carriers" of DNA. DNA not only determines what species an organism will be, but many other characteristics from number of limbs to whether it will be a mammal or not. Humans are the focus of most genetic research, as heredity may hold the key to curing cancer and other diseases.

Geneticists act like medical detectives who hunt for clues to explain how genes are expressed and regulated in the cell and how genes are copied and passed on to successive generations.

1. INTRODUCE

About Mary Styles Harris: Mary Styles Harris earned a Ph.D. in genetics from Cornell University in 1975.

Ask students what kinds of things they think Dr. Harris might have liked as a child. Point out that researching genes is a lot like putting puzzles together.

About Harris' Work: Dr. Harris performed research and then served as executive director of the Sickle Cell Foundation of Georgia and later as the first director of Genetic Services for the state. Sickle cell anemia is a disease that can be passed down in families. Dr. Harris also hosted a series of television and radio programs about issues African Americans are facing. Today she is the executive producer of "Journey to Wellness," a syndicated radio program and newspaper column. Dr. Harris is president of her own company, Biotechnical Communications, Inc., in Atlanta, Georgia.

2. DISCUSS

Use the copy master to prompt discussion. Lead students to understand that the study of genes can help scientists understand diseases and make cures and treatment for them.

3. EXTEND

Share with students some of the information about genetics presented in Science Background. Students might like to share similar genetic characteristics that exist in their own families.

reading aloud

Tim's class is drawing family trees. They are looking at the number of people in each student's family who have the same eye and hair color. Tim's teacher explains that we each have our own unique features—no two people are exactly alike. Certain groups of people, however, share some similar characteristics. Families have similar characteristics. This is due to heredity: If your parents and grandparents had blue eyes, you probably will too. Some characteristics, like hair and eye color, are passed from family member to family member. Abilities can also be passed down, such as singing or playing sports.

Mary Styles Harris' interest in science came from her father, who was a doctor. Harris liked to ask him questions and do research with him. She liked it so much that she would enter science fairs at school and volunteered at the first black-owned medical laboratory. Harris learned that diseases like diabetes and sickle cell anemia can be passed in families. She wanted to learn more.

Kelly Miller

The picture shows Kelly Miller looking at the night sky.
Draw a circle around the stars in the sky.
How many stars are there?
Color the picture.

Kelly **Miller**
TEACHING NOTES

Purpose: To introduce students to the first African-American mathematics graduate student.

Science Background: Many scientists use math to help them study astronomy. Astronomy is the study of the universe, including planets and stars.

Much of what we know about our universe is from using mathematics to find out how far the other planets and stars are from Earth, how big they are, and how fast they travel.

I. INTRODUCE

About Kelly Miller: Kelly Miller was born on July 18, 1863. His education in mathematics began when Reverend Willard Richardson recognized that Miller was talented and recommended that he be allowed to go to college. Miller was awarded a scholarship to attend Howard University, where he studied mathematics, Latin, and Greek. He became the first African American to be admitted to Johns Hopkins University. He studied advanced mathematics and astronomy. Miller went on to receive a master's degree in mathematics and a law degree.

About Miller's Work: When an increase in tuition prevented Miller from continuing his studies, he left Johns Hopkins and began a career in teaching. In 1907 he became dean of the College of Arts and Sciences at Howard University. In addition to writing numerous articles and essays, he had a weekly newspaper column. Miller was an outspoken advocate for African Americans obtaining college degrees. His articles sparked much controversy. He died on December 29, 1939.

2. DISCUSS

Use the copy master to prompt discussion. Lead students to discuss why math is an important skill to learn.

3. EXTEND

Ask students to share their pictures with the class.

reading aloud

Frank and his father went outside to look at the stars in the night sky. Frank started to count the stars and counted 20! Some stars were bigger than other stars.

When Frank was counting the stars in the sky and comparing their sizes, he was using math skills. Kelly Miller was a mathematician who used his math skills to study the stars and planets in the sky.

Even as a young boy, Miller was always good at math. Because of his skills, he was allowed to go to a college that specialized in math. After studying math and astronomy in college, Miller became a teacher. He worked hard to teach math and other subjects to African Americans. Miller tried to develop opportunities for African Americans to go to college and study math and sciences. He wrote many articles and essays on this and the rights of African Americans.

Norbert Rillieux

A long time ago, a man named Norbert Rillieux
changed the way sugar looks and tastes.
This picture shows Norbert Rillieux next to some sugar
cane plants. How many is he holding?
Circle one plant.
Color the picture.

Norbert Rillieux
TEACHING NOTES

Purpose: To introduce students to an inventor noted for his contribution to the process of refining sugar.

Science Background: The "Rillieux (rihl YOO) System" evaporator works on the principle that as pressure increases, solutions cook faster and at a lower temperature.

Refining sugar is a two-step process: 1) The clarified juice is extracted from sugar cane, and 2) the water in the juice must be evaporated by boiling. Rillieux developed a way of evaporating the water from the juice. He did this in a connected series of closed pans. First, he removed all or part of the air in the pans to lower the pressure. Then, he brought the juice in the first pan to the boiling point. He used the steam from each pan to boil the juice in the next pan. The result was that each pan operated under greater pressure than the one before it, requiring less and less heat to boil the sugar juice.

Rillieux's evaporator saved fuel because each pan used the heat of the pans before it in the process. Less heat was wasted, so more fuel was saved. The evaporator also made the process safer, faster, and less expensive.

I. INTRODUCE

About Norbert Rillieux: Rillieux was born in New Orleans, Louisiana, in 1806. He was the son of a slave and her wealthy French owner. Rillieux's father sent him to study engineering at L'Ecole Centrale in Paris, where he specialized in steam engine technology.

About Rillieux's Work: Rillieux invented a process called *multiple-effect evaporation*, which changed the way sugar was made from sugar cane. Today, the process is used to manufacture not only sugar, but also soap, glue, gelatin, and condensed milk. The "Rillieux System" was so successful that many other countries that produced sugar cane, such as Mexico and Cuba, wanted to use it.

2. DISCUSS

Use the copy master to prompt discussion. Lead students to discuss the ways we use sugar.

3. EXTEND

Discuss other products that are manufactured by Rillieux's process, such as soap, glue, and gelatin. How are these products similar? How are they different? In what ways are these products used today?

‹‹ reading aloud

Do you like sugar? Today, sugar is easy to get. But many years ago, sugar was a special treat that only a few people could afford. Sugar was expensive because it took a very long time to make. Making sugar was also very dangerous. That's because in order to make sugar, sugar cane (the plant that sugar comes from) had to be cooked. The sugar cane was cut into pieces and boiled in big pots. Workers poured the boiling sugar cane juice from one pot to another. Many workers would burn themselves with the hot liquid. After cooking it for a long time, the juice got thick. When it turned brown and lumpy, the workers stopped cooking it. That was sugar.

Because Norbert Rillieux knew that the process of turning sugar cane into sugar, called refining, was a slow, expensive, and dangerous process, he decided to find a better way. He thought of a new way to get sugar from sugar cane that was faster, less expensive, and safer. Also, his invention made the sugar into white crystals, not brown lumps like before. Soon, many sugar plantations were using Rillieux's refining system.

Lawnie **Taylor**

The picture shows Dr. Taylor looking at solar panels.
Draw a circle around the sun.
Color the picture.

Lawnie Taylor
TEACHING NOTES

Purpose: To introduce students to a physicist who studied solar technology.

Science Background: The sun, Earth's closest star, feels hot and dominates the sky because it is 250,000 times closer than the next closest star to Earth.

The center, or core, of the sun is very hot. A process called nuclear fusion takes place in the sun's core. Nuclear fusion produces a lot of energy. Some of this energy travels out into space as heat and light, and some of it hits Earth.

Scientists like Dr. Taylor try to capture the sun's energy and use it here on Earth to heat homes, grow foods, and make machines run.

I. INTRODUCE

About Lawnie Taylor: Dr. Lawnie Taylor was born in 1920. He graduated from Columbia University with a degree in physics. He completed his doctoral work at the University of Southern California.

About Taylor's Work: For 27 years, Dr. Taylor was senior engineer for the U.S. Department of Energy. He worked on nuclear waste cleanup, solar energy research, and technology information exchange, among other projects. Before that he was a program manager for the Energy, Research, and Development Administration and a senior physicist for Xerox Corporation. He retired from government work in 2002.

Dr. Taylor owns his own company, LHTaylor Associates. Point out to students that as a physicist, Dr. Taylor uses many skills such as problem solving, creative thinking, and observation. Ask students what they think Dr. Taylor enjoyed learning in school.

2. DISCUSS

Use the copy master to prompt discussion. Lead students to discuss how physical scientists study objects in their physical world, such as the sun, water, and wind.

3. EXTEND

Have students share their drawings with the class.

◄◄ reading aloud

Today at school, Francine's class did an experiment. They observed the effect of the sun's heat on different things. They placed a crayon, an ice cube, and a pencil on a sheet of a paper and put the paper in the sun. After an hour, they looked at the paper to see what happened. Francine saw that the crayon had melted. So did the ice cube. The pencil didn't melt, but it felt hot.

Exploring the effect of the sun's heat is an example of physics, a science that deals with matter and energy. Dr. Lawnie Taylor is a physicist. He studied how to use the sun's energy to make homes warm, grow food, and make machines run. While at NASA, he worked on experiments about rockets.

Dr. Taylor liked to study physics because he felt it was a way to make a difference. He is still trying to make a difference. In 2004, he invented a product to remove stains from cotton without ruining the fabric.

George Washington Carver

AT A GLANCE

lived 1860–1943
Diamond Grove, Missouri

education
B.S., M.S., Iowa State College
of Agriculture and Mechanic Arts
(Iowa State University)

occupation
Scientist, Educator

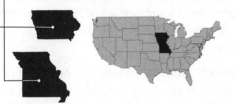

Do you ever wonder what makes plants grow? George Washington Carver did. When he was a young boy, he loved to work in the garden and watch things grow.

Carver became a scientist. He studied plants and the kind of soil they grew in. Carver planted crops to see which ones made the soil better. Cotton was not good for the soil. But peanuts made the soil richer. He told farmers about his discovery. Then he invented 325 ways to use peanuts.

● **CARVER** in the lab

? explain it

1. How did George Washington Carver help farmers?
2. What did cotton do to the soil? What helped the soil?

! explore it

Find out what other things Carver invented that are made from peanuts.

George W. Carver
TEACHING NOTES

Purpose: To introduce students to an American scientist whose work enabled farmers to increase their crop yields and whose inventive nature helped our country.

Science Background: Crop rotation is a proven method for increasing crop yield. Crops such as tobacco and cotton, which leach nutrients from the soil, are rotated with crops such as peanuts and soybeans, which rebuild the soil.

Peanuts and soybeans are examples of legumes. Legumes are plants that have nodules of nitrogen-fixing bacteria on their roots. These bacteria take the nitrogen from the air and incorporate it into the soil. Nitrogen is a necessary nutrient for plant growth.

Botany is the study of plants. A scientist who studies plants is called a botanist. As a botanist, Carver used plants such as the sweet potato to make products that were commonly used. From the sweet potato, he discovered 118 products that could be used as substitutes for items that were hard to find during World War I.

1. INTRODUCE

About George Washington Carver: Carver was a scientist who studied plants. He taught others how to make things from nature and how to help plants grow better. Ask students, "How many different things can you make out of a peanut?" Count how many different ideas students had. Explain that George Washington Carver was a scientist who came up with 325 uses for a peanut. One of them was something many of us eat for lunch: peanut butter!

About Carver's work: Ask students what roles curiosity, enthusiasm, and creativity played in Carver's work. How does asking questions and investigating things make them like scientists?

2. DISCUSS

Use the copy master to prompt discussion. Lead students to understand that because of Carver's discovery, he had to use problem solving skills to investigate ways peanuts could be used. Why is problem-solving an important skill for scientists to know and use?

Answers to Explain It:
1. He helped them to make the soil better so they could grow more cotton.

2. Cotton was bad for the soil. It took nutrients out of the soil. Peanuts helped the soil.

3. EXTEND

After students have completed the Explore It, ask them to share what they have learned. Ask students, "Where are these things used? At home? At school?"

Career Focus: A botanist learns about plant life. What kinds of plants can hurt us? Help us?

◄◄ reading aloud

When George Washington Carver was a young boy, he was very sick and spent a lot of time in the garden and walking in the woods. He became interested in plants. At age 30, Carver went to college to learn about plants. He was the first African American to graduate from his school. Because he enjoyed learning, he became a teacher. He was also a scientist.

One thing that Carver studied was soil. The soil in Alabama was very poor. Carver thought that changing the plants that were grown in the fields would help the soil. One of the major crops grown in Alabama is cotton. Cotton is not good for the soil because it uses up many of the things in the soil that plants need to grow. Carver found that planting peanuts helped the soil become better for other plants. He told farmers to plant peanuts one year and cotton the next year. That process is called *crop rotation*. When they did this, the farmers were able to grow more cotton.

Gloria Ford Gilmer

AT A GLANCE

education
B.S. Morgan State University
M.A., University of Pennsylvania
Ph.D., Marquette University

occupation
Mathematician

Gloria Ford Gilmer loves math. She also loves teaching math.

Dr. Gilmer became interested in a type of math called ethnomathematics. It is the study of how people from different groups do math. She helped form an international study group in ethnomathematics.

Can you see math patterns in hairstyles?

Today, Dr. Gilmer is president of Math-Tech. Her company develops math programs for women and minorities.

? explain it

1. What is ethnomathematics?
2. What group did Dr. Gilmer help to create?
3. What does Dr. Gilmer do today?

! explore it

Dr. Gilmer recently studied math patterns in African-American hairstyles. Look at the hairstyles in your class. What patterns can you find?

Gloria Ford **Gilmer**
TEACHING NOTES

Purpose: To introduce students to a mathematician who became the first woman to give the National Association of Mathematicians' Cox-Talbot Address.

Science Background: Ethnomathematics is the study of how mathematics is used in different cultural groups. It was created when researchers discovered evidence that different population groups, or people from different societies, performed math in ways that are not necessarily like those in Western societies.

Mathematicians like Dr. Gilmer study ethnomathematics to have a greater understanding of the cultural diversity in how math is performed and how to apply this understanding to develop math education programs.

I. INTRODUCE

About Gloria Ford Gilmer: After earning an M.A. in mathematics from the University of Pennsylvania and teaching and raising her own children, Gloria Ford Gilmer went back to school for her Ph.D. Dr. Gilmer earned her Ph.D. in curriculum and instruction from Marquette University.

About Gilmer's Work: In 1980-1982, Dr. Gilmer was the first African-American female to be on the board of governors of the National Associa-

tion of Mathematicians (NAM). In addition, she served as a research associate with the U.S. Department of Education. In 1992, Dr. Gilmer had the honor of being the first woman to give the NAM's Cox-Talbot Address. When Dr. Gilmer became interested in the field of enthomathematics, she helped form the International Study Group on Ethnomathematics. Today, Dr. Gilmer is president of Math-Tech. Her company uses research findings to develop math programs.

2. DISCUSS

Use the copy master to prompt discussion.
Talk about some of the traits that Dr. Gilmer must have had to become a teacher, researcher, and writer of mathematics.

Answers Explain It:
1. study of how people from different groups perform mathematics

2. She is co-founder of an international study group on ethnomathematics.

3. She is president of Math-Tech, a company that develops mathematical programs for women and minorities.

3. EXTEND

After students have completed the Explore It, ask them to share what they have learned. Why is helping women and minorities learn math important to Dr. Gilmer?

Career Focus: Are you interested in becoming a mathematician? Why or why not?

◄◄ reading aloud

"Wow!" Tracey thought. "I got an A on that math test." Tracey really likes math. It is neat to figure things out. Because she is so good at math, her teacher asked her to help some of her classmates. Tracey likes to help her friends learn how to add and subtract numbers. It feels good.

Like Tracey, Gloria Ford Gilmer likes to help others learn about math. She became a researcher in mathematics and wanted to teach others about her research findings. Today, Dr. Gilmer is the president of a company that translates research findings into useful mathematics education programs.

Dr. Gilmer studies how people from different groups do math. She has become a leader in the field of ethnomathematics. Ethnomathematicians study how other people, from all around the world, learn and use math.

Mae C. Jemison

AT A GLANCE

- **born** 1956
  Decatur, Alabama
- **education**
  B.S., B.A., Stanford University
  M.D., Cornell University
- **occupation**
  Physician, Astronaut

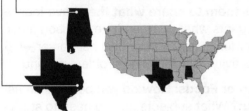

What do you want to be when you grow up?

When Dr. Mae Jemison was a little girl, she wanted to become a scientist.

After Jemison went to college, she studied medicine. She became a doctor. Then she decided to become an astronaut.

● **DR. JEMISON** conducting experiments in space

In September of 1992, her dream came true. She flew a space shuttle mission with six other astronauts. In space, Dr. Jemison studied ways to keep astronauts from getting sick.

? explain it

1. What does an astronaut do?
2. What did Dr. Jemison study in space?

! explore it

Go to the media center at your school. Ask for help to find out how people become astronauts.

Mae C. Jemison
TEACHING NOTES

Purpose: To introduce children to a scientist and physician who was the first African-American woman to fly in a space shuttle mission.

Science Background: Motion sickness affects astronauts in space, just as often as it affects people on Earth. People get motion sickness when their sense of balance is upset. It can happen when they ride in a car or on a ride at an amusement park.

Our sense of balance is centered in the inner ear, in a section called the *vestibule*. The vestibule is divided into two sections, the *saccule* and the *utricle*. These sections are lined with hair cells, which have crystals of calcium carbonate stuck to their outer surface. When gravity pulls the crystals down, they create nerve impulses that are sent to the brain. The stronger the pull on the crystals, the stronger the nerve impulses. These impulses are interpreted in the brain and tell the person which way is up and which way is down. In space there is no up or down, so an astronaut's sense of balance is upset. This can make some astronauts feel sick.

I. INTRODUCE

About Mae C. Jemison: When Dr. Mae Jemison was 5 years old, her kindergarten teacher asked her what she wanted to be when she grew up.

Mae said, "I want to be a scientist." Dr. Jemison became the first African-American woman to go into space.

About Jemison's Work: Dr. Jemison worked inside Spacelab J, a pressurized laboratory located in the cargo bay of the space shuttle. The crew lived and worked inside the lab and conducted 43 major experiments for eight days. Dr. Jemison founded and now chairs The Earth We Share, an annual space camp where students work together to solve current global problems.

2. DISCUSS

Use the copy master to prompt discussion. Lead students to understand how much astronauts need to learn. For example, in addition to her medical training, Dr. Jemison needed courses in wilderness, water, and parachute survival.

Answers to Explain It:
1. An astronaut explores space and performs experiments.

2. She studied ways to keep astronauts from getting sick in space.

3. EXTEND

After students have completed the Explore It, ask them to share what they have learned. Ask students what qualities Mae Jemison must have in order to be successful in her career—e.g., creativity, curiosity, and problem-solving skills.

Career Focus: How do you become an astronaut? What subjects do you need to study?

<< reading aloud

Latisha loved riding on the roller coaster. It was her favorite ride at the amusement park. She liked how it went up and down and when it went upside down. Her brother, on the other hand, didn't like riding on the roller coaster because it made him feel sick. His stomach would start to ache, and he would feel like he was floating. Her brother had *motion sickness*. This means that when people move in certain ways, they feel sick.

Some astronauts get motion sickness when they travel in space. When Dr. Jemison was on the space shuttle *Endeavour* on September 12, 1992, she looked for ways to keep astronauts from getting motion sickness. She tried a process called biofeedback. When people can control their breathing, heart rate, and skin temperature, they can keep from having motion sickness. It is important for astronauts to avoid becoming sick in space because they have a lot of work to do.

Lewis H. Latimer

AT A GLANCE

lived 1848–1928
Chelsea, Massachusetts

education
worked with Thomas Edison

contribution
Artist, Inventor; invented a way to
make the lightbulb last longer

Do you like to draw and make new things? Lewis H. Latimer did. He was an artist and inventor.

When he was young, people asked him to draw pictures of their inventions. They used the drawings to get a *patent* for their invention.

A patent is a license. It says that only that person has the right to sell it.

LATIMER'S lightbulb

Latimer invented a way to make the tiny wires inside of a lightbulb better. These new wires helped make the lightbulb last longer.

? explain it

1. How did Lewis Latimer help others?
2. How did Latimer's invention improve the lightbulb?

! explore it

Go to the media center at your school and ask for help to find out more about Lewis Latimer and his inventions.

Lewis H. Latimer
TEACHING NOTES

Purpose: To introduce students to a self-taught scientist who was also an artist and an inventor.

Science Background: The incandescent lightbulb is an invention that has changed the way we live. The most difficult part about perfecting the lightbulb was making the filament. If you look inside a lightbulb, you can see very thin wires that connect two thicker wires. When these wires heat up, they glow, producing the light you see.

During the process of inventing the incandescent lightbulb, many different substances were used to try to make the filaments. None of them lasted long enough to make the lightbulb practical. Then Latimer discovered how to coat the thin thread with carbon to make it strong enough to withstand the heat produced inside the bulb.

I. INTRODUCE

About Lewis Howard Latimer: Ask students why lightbulbs were a good invention. Explain that the first lightbulb didn't glow for very long—only a little more than 40 hours. Lewis H. Latimer, a scientist, wanted to find a way to make a lightbulb last longer. After many experiments, he was successful.

About Latimer's Work: Latimer worked in the field of electricity. He worked with Thomas Edison for many years and as a draftsman for different companies. Latimer wrote a book on electric lighting. He also helped install some of the first incandescent electric light stations in New York City.

2. DISCUSS

Use the copy master to prompt discussion. Talk with students about what it means to be an inventor. Ask students how they think curiosity, creativity, openness, and experimenting were important in Latimer's work.

Answers to Explain It:
1. He drew pictures of their inventions. This helped them get a patent.

2. It made the lightbulb last longer.

3. EXTEND

After students have completed the Explore It, ask them to share what they have learned. Have students choose one of the inventions they learned about and create a drawing of it. Have them write a small explanation of the invention and how the invention may be used today.

Career Focus: People who like to draw and design sometimes choose a career in drafting. What do you think draftspeople do?

reading aloud

Lewis H. Latimer was an artist and an inventor. When he was 10, he got a job as an office helper for a group of lawyers. These lawyers worked with people who applied for patents for things they had invented.

Every patent application had to have drawings of the invention. Latimer bought a used set of drawing tools and began to learn all he could about the job. One day the lawyers said he could start making drawings for inventions.

One of his first drawings was for Alexander Graham Bell for a patent on the telephone. Next, Latimer began working on some inventions of his own. His most famous patent was for a new way to make the little wires you see inside a lightbulb last longer.

Alexander Miles

AT A GLANCE

born
Duluth, Minnesota

education
Informal

occupation
Inventor

Do you like to go up and down in an elevator? Elevators can be fun. But they were not always safe.

You can thank Alexander Miles for that smooth, safe ride. He helped improve elevators.

The idea of an elevator goes back to ancient times. Early Egyptians built giant pyramids. They used a hoist to help lift the heavy stones.

○ Old-fashioned elevators had operators.

Alexander Miles invented an electric elevator with safety features. The doors easily opened and closed.

? explain it

1. Who was responsible for improving elevators over the years?
2. Some modern elevators are "smart." What do you think this means?

! explore it

The ancient Egyptians possibly used an early form of the elevator to build the Great Pyramid of Cheops at Giza. Collect some interesting facts to share with the class.

Alexander Miles

TEACHING NOTES

Purpose: To introduce students to an inventor who helped improve the quality and safety of elevators.

Science Background: The earliest elevators were simple lifting devices called hoists. They were powered either by animals or by people. The ancient Egyptians may have used a primitive hoist to create the 481-foot tall Great Pyramid at Giza. The pyramid was built with stones weighing up to five tons each. Hoists were also utilized during the Middle Ages. On the French coast, the Abbey of Mont St. Michel had a treadmill hoisting machine installed in 1203. The machine was powered by donkeys.

The first electrically powered elevator was introduced at the 1889 Paris Exhibition. During that same year, Elisha Otis produced the first electric elevator in New York City. Improvements in the design of Otis' elevator paved the way in the early 1900s for the building of skyscrapers.

1. INTRODUCE

About Alexander Miles: Not much is known about the personal life of Alexander Miles.

About Miles' Work: Miles received a patent in 1887 for an electric elevator. His elevator incorporated several improvements over earlier models. It had a mechanism that greatly improved the ease with which the cab doors were opened and closed. In addition, it had a mechanism that automatically closed the shaft door.

2. DISCUSS

Use the copy master to prompt discussion. Emphasize to students that many inventions are not the result of one person's efforts. Rather, they are the result of many people's efforts over long periods of time and in many locations.

Answers to Explain It:

1. Many people, including Alexander Miles, were responsible for improved elevators.

2. Answers will vary. Possible answer: Some modern elevators are probably controlled by computers.

3. EXTEND

After students have completed the Explore It, ask them to share what they have learned. Some students might like to draw the hoisting mechanism used by the ancient Egyptians to build the Great Pyramid. Have them write step-by-step captions that show how the pyramid was built.

Career Focus: Think about ways you could improve modern elevators.

<< reading aloud

Marsha was so excited. She was going on vacation with her family. They were staying at a hotel with a glass elevator. She loved riding up and down the elevator. She could see almost the whole city when she rode up to the top floor.

Elevators have come a long way since the hoists of ancient times. In the Middle Ages, donkey power helped run a treadmill hoisting machine. In the 1800s, they had steam-powered elevators. Many people helped develop the modern elevator. Elisha Otis made the first passenger elevator. Improvements in his design paved the way for building skyscrapers. Alexander Miles invented an electric elevator at about the same time. He developed a shaft door that closed automatically. The older models had to be closed by hand. Elevator operators sometimes forgot to close the door, causing serious injuries.

Name

Name

Date

Joseph **Pierce**

AT A GLANCE

lived 1902–1969
Waycross, Georgia

education
A.B., Atlanta University
M.S., Ph.D., University of Michigan

occupation
Mathematician, Professor

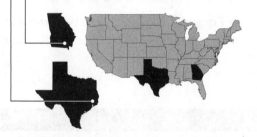

How many different ways can you group the students in your class?

Joseph Pierce liked questions like this. He used math to answer them.

Dr. Pierce became a mathematician. He found a formula that could be used to study populations, or groups of people.

PIERCE used math to study groups of people

Dr. Pierce also helped groups of people, such as African Americans. He helped them learn skills, such as math, so that they could find good jobs.

? explain it

1. What is a population?
2. What did Dr. Pierce use to help him answer questions?
3. How did Dr. Pierce help African Americans?

! explore it

Find out more about other groups or people that you belong to. Put your information on a graph. What does this graph tell you?

Joseph A. Pierce
TEACHING NOTES

Purpose: To introduce students to a mathematician who made a contribution to the field of statistics.

Mathematics Background: Mathematicians are people who use math to help them solve problems and understand the world around them. One type of tool that mathematicians use is called statistics. Statistics is the collection, organization, presentation, interpretation, and analysis of data. Statistic is a word for any kind of measurement or count.

I. INTRODUCE

About Joseph Pierce: Joseph Alphonso Pierce enjoyed teaching mathematics. He first became interested in math after college, when he accepted an assignment as assistant coach at Texas College in Tyler, Texas. When he arrived at the school, he learned that he would be required to teach mathematics. After teaching math for four years, Dr. Pierce liked it so much that he decided that math would become his profession. Dr. Pierce earned his M.S. and Ph.D. in mathematics from the University of Michigan.

About Pierce's Work: Dr. Pierce was a professor of mathematics at Wiley College in Texas. In 1952 he was named dean of graduate studies at Texas Southern University (TSU). He eventually became the president of TSU.

Ask students what roles curiosity, creativity, and problem solving played in Dr. Pierce's work. How do students use these skills when they study math?

2. DISCUSS

Use the copy master to prompt discussion. Lead students to understand that in order for Dr. Pierce to create his formula for grouping data together, he had to use problem-solving skills. Why is problem solving an important skill for mathematicians to know and use?

Answers to Explain It:

1. A population is a group of people or things.

2. Dr. Pierce used mathematics.

3. Dr. Pierce helped them to learn skills and find jobs.

3. EXTEND

After students have completed the Explore It, ask them to share what they have learned. Ask students to investigate and graph other groups that they want to learn more about.

Career Focus: If you wanted to become a mathematician, what types of classes would you need to take?

<< reading aloud

What letter does your name start with? Does anyone else's name in your class begin with this letter? What is the most common name in your school? What is the least common? By answering questions like these, you are using statistics. Statistics is a way to count things.

Dr. Joseph Pierce was a mathematician who used statistics to help him group information. By doing so, he could find out how many times or how few times something occurred. One important way he used statistics was to help him find the number of jobs that were available to African Americans. He counted the number of jobs available and compared that to the number of jobs that weren't. He found that there were only a small number of jobs available to African Americans.

Dr. Pierce wanted to know why. He found that one reason was because many African Americans weren't being taught certain skills, such as math. He then tried to help African Americans learn math and other important skills so that they could find jobs.

Lisa D. White

AT A GLANCE

- **born** 1961
  East Lansing, Michigan
- **education**
  B.A., San Francisco State University
  Ph.D., University of California
- **occupation**
  Micropaleontologist

Do you enjoy puzzles? If you do, you might want to become a paleontologist. Some paleontologists study fossils that are very small. The fossils are about as wide as the edge of a dime.

Dr. Lisa White is a scientist who studies small fossils. She learns about

○ **DR. WHITE** looks for fossils in Zimbabwe.

events that happened long ago. She solves puzzles about the Earth. She also learns about what Earth was like when dinosaurs lived.

Today, Dr. White is a teacher.

? explain it

1. What does a micropaleontologist do?
2. What can micropaleontologists find out from fossils?
3. What does Dr. White teach students?

! explore it

Use the Internet to find the difference between a paleontologist and a micropaleontologist.

Lisa D. White
TEACHING NOTES

About White's Work: Dr. White is professor of geology and chair of the Geosciences Department at San Francisco State University. In her work, Dr. White travels to many places around the world to study fossils. She was awarded a grant by the National Science Foundation to help students develop an interest in a career in earth science.

Purpose: To introduce students to a micro-paleontologist and university professor.

Science Background: Certain types of microfossils are used as indicators for petroleum deposits. So, when *petrologists* (geologists who search for petroleum) find an area that is likely to have petroleum, they insert a hollow tube into the rock and pull out a thin sample. Each layer of rock is then examined for signs of microfossils. Another way to find microfossils is by examining indicator rock. Large numbers of microfossils are found only in certain types of rock. When deposits of these rocks are found, micropaleontologists examine the rock for microfossils.

Most microfossils are too small to be seen with the unaided eye. Micropaleontologists use microscopes and other magnifying equipment to examine the fossils.

1. INTRODUCE

About Lisa White: Dr. Lisa White is a micropa-leontologist—a scientist who studies very small fossils. While on summer break from college, White worked with a geologist (a scientist who studies Earth) who was also a micropaleontolo-gist. It was then that she knew she wanted to become a micropaleontologist herself.

2. DISCUSS

Use the copy master to prompt discussion. Lead students to understand that it is through the study of fossils that we know about life on Earth long ago.

Answers to Explain It:
1. A micropaleontologist studies fossils that are very small.

2. Answers will vary. Possible answer: They find out how long ago things happened and what Earth was like when dinosaurs lived.

3. Dr. White teaches students they can learn about Earth by studying fossils.

3. EXTEND

After students have completed the Explore It, ask them to share what they have learned. Have students find items in the classroom that are less than 1 mm in size and then measure them, using a ruler.

Career Focus: If you were interested in becoming a micropaleontologist, what classes would you need to take?

<< reading aloud

"Look what I found!" shouted Maria as she pulled her mother over to a large rock. There in the layers of the rock was a very, very tiny imprint of a plant. It was about as big as a dime. "I wonder how old it is?" asked Maria. "Your aunt might be able to tell you," said Maria's mother. "She is a paleontologist. She studies fossils like these as part of her job."

As a micropaleontologist, Dr. Lisa White studies fossils that are less than 1 mm in size. These fossils are the remains of single-celled plants and animals that lived millions of years ago.

Dr. White was born in East Lansing, Michigan. As a little girl she was very shy, but she enjoyed spending time with her two older sisters listening to music, roller skating, and dancing. White's family moved to California. At school, Dr. White enjoyed learning about nature.

Today, Dr. White likes to work with scientists from other countries. She also enjoys her job as a teacher because she can share with her students what she has learned.

Paul **Williams**

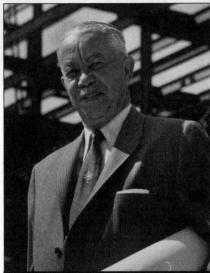

AT A GLANCE

lived 1894–1980
Los Angeles, CA

education
Los Angeles School of Art
Beaux Arts School of Design
University of Southern California

occupation
Architect

Paul Williams was an architect. Architects design buildings. They have to know math and engineering.

Williams first worked as a landscape architect. He designed the gardens that surround buildings.

Then he learned to design buildings that fit in with their surroundings.

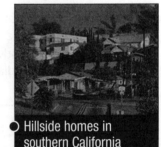

○ Hillside homes in southern California

People liked Williams' designs. His work became popular around the world.

Today, many people still use his designs.

? explain it

1. What does an architect do?
2. What was Paul Williams' first job?
3. What did Williams learn at his first job?

! explore it

What is the difference between an architect and a landscape architect?

Paul Williams
TEACHING NOTES

Purpose: To introduce students to an architect who used his skills to build public buildings and private homes and design landscaping.

Science Background: The goal of a landscape architect is to beautify the places where people live and work.

Landscape architects use computers, computer-aided design (CAD) tools, and video simulation equipment to design livable environments and make detailed plans. Architects use science and math to make buildings and surroundings that are both interesting and functional.

Landscape architects work on projects such as deciding on the best type of shade tree to use in a yard, or they can work on complex projects such as restoring a woodland preserve.

I. INTRODUCE

About Paul Williams: Paul Williams loved art and was interested in design. After high school, he went to school to study these things. That is when he took his first job as a landscape architect. Williams also designed buildings.

About Williams' Work: Williams designed over 2,000 buildings, from mansions for celebrities to public buildings such as Los Angeles International Airport. His career lasted over half a century. Williams received many awards for his work. He also published two books: *Small Homes of Tomorrow* and *New Homes for Today*.

2. DISCUSS

Use the copy master to prompt discussion. Lead students to an understanding of building and design by doing this activity. In groups, have students choose one of their favorite places in the community and draw a picture of it. Students should write a paragraph describing how it was designed so that it fit in with its surroundings. Have them include landscaping. Allow groups to share their drawings with the class.

Answers to Explain It:

1. An architect designs buildings.

2. He was a landscape architect.

3. He learned the importance of designing buildings that fit their surroundings.

3. EXTEND

After students have completed the Explore It, ask them to share what they have learned. Discuss how the two jobs work together.

Career Focus: If you are interested in becoming an architect, Paul Williams would probably have told you to study art, ecology, biology, and math in school.

<< reading aloud

Brad and Michele are excited! They are going to the new skate park after school today. Their father is a landscape architect and he helped to build it. It makes their neighborhood look much prettier. Before it became the park, the area used to be an old garbage dump. Brad and Michele's father designed the plans that changed the ugly dump into a beautiful new park where they could play.

Like Brad and Michele's father, Paul Williams worked to make the space where people live and work more beautiful. He designed buildings that looked good in the places where they were built. To do that, he had to learn about different ways to construct buldings. He also had to learn more about the environment and plants. Because of the the wonderful work that he did, Williams received honorary doctorate degrees from Howard University, Lincoln University, and Tuskegee University.

James E. Bowman, Jr

AT A GLANCE

born 1923
Washington, D.C.

education
B.S., M.D., Howard University

occupation
Physician, Geneticist

Why do you have brown eyes or straight hair? It is your genes. A gene is part of a cell. You get your genes from your parents.

Dr. James Bowman studied how genes help decide what we look like. He also studied genes that cause diseases like sickle cell anemia. Sickle cell anemia is a blood disease that is passed from parents to children.

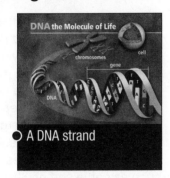

A DNA strand

Dr. Bowman thinks that someday science will conquer this disease. But he worries about science grouping people based on their genes. However, he knows that scientists can solve this problem.

explain it

1. How is science conquering diseases?
2. Why does Dr. Bowman say that genetic science should be used wisely?

explore it

Who was Gregor Mendel? Use the Internet to find out why he is important in the field of genetic science.

James E. Bowman, Jr.
TEACHING NOTES

Purpose: To introduce students to a physician who has done important research with genetic diseases, such as sickle cell anemia.

Science Background: Genetics is the study of heredity. Heredity is the process by which parents pass traits on to their offspring. Genes determine physical traits such as eye and hair color and physical build. Some diseases, such as hemophilia and sickle cell anemia, are also passed on genetically. Geneticists study the passing on of traits through genes.

People have long used their knowledge of genetics to improve crops and livestock. In the 1970s and 1980s scientists developed genetic engineering by splitting DNA molecules and splicing them together to make a new hybrid molecule. This molecule is called recombinant DNA. It can be inserted into an organism to change the organism's physical traits. Genetic engineering has improved many types of crops and animals such as cattle and poultry.

In the last few decades, tests have been developed that indicate whether a person carries genes for particular diseases that could be transmitted to offspring. Genetic counselors help people understand their genetic makeup through tests like these and family history.

I. INTRODUCE

About James E. Bowman, Jr.: James Bowman was born in Washington, D.C., in 1923. Dr. Bowman earned a B.S. and an M.D. from Howard University.

About Bowman's Work: Dr. James Bowman is professor emeritus at the University of Chicago in the pathology and medicine departments. He serves on the Committee on Genetics and the Committee on African and African-American

Studies and is a scholar for the MacLean Center for Clinical Medical Ethics. Dr. Bowman is dedicated to using genetic science wisely to understand traits and diseases that occur in particular populations. He warns against limiting groups of people by identifying their genetic makeup. He says, "Genetic variation is used to delineate populations and ethnic groups," but "often there is more *intra*group variance than *inter*group variance." The continent of Africa is a prime example. It is home to the tallest group of people in the world—the Tutsis, and the shortest—the pygmies.

2. DISCUSS

Use the copy master to prompt discussion. Have students give examples of inheriting eye color, hair color, and so on from their parents.

Answers to Explain It:
1. Answers will vary. Possible answer: Examples include the development of medicines that cure diseases, vaccines that prevent diseases, new ways of diagnosing diseases so that they may be treated early, and new surgeries and other techniques such as radiation that cure diseases.

2. Answers will vary. Possible answer: Genetics is concerned with the traits that individuals and groups of people inherit. Grouping people based on their physical traits points out differences between groups of people. It may cause people to stereotype groups or even mistreat them based on their traits.

3. EXTEND

After students have completed the Explore It, ask them to share what they have learned. Mendel discovered the principles of genetics in the 1860s through his study of the traits that garden peas inherited.

Career Focus: Would you be interested in studying genes? Why, or why not?

Alexa Canady

AT A GLANCE

born 1950
Lansing, Michigan

education
B.S., M.D., University of Michigan

occupation
Neurosurgeon

Suppose you were looking for Dr. Alexa Canady at Detroit's Children's Hospital a few years ago. You might have found her doing brain surgery or playing video games with a patient. "I like to play, so my day always includes it," she said. That's one reason Dr. Canady chose to work with children.

○ **CANADY** takes a break from surgery.

Dr. Canady was the first African-American woman to become a brain surgeon. She performed 12 operations a week. But she always made sure she knew her patients. "It is so important that patients are able to talk to you," says Dr. Canady.

? explain it

1. Do you think playing video games would be good for a child in the hospital? Explain why or why not.
2. Why did Dr. Canady want her patients to be able to talk to her?

! explore it

Some children need brain operations after getting hurt. Find out some ways to protect your head from injury.

Alexa **Canady**
TEACHING NOTES

Purpose: To introduce students to the first African-American woman to become a neurosurgeon.

Science Background: College students who plan to enter medical school usually take a liberal arts program that includes advanced math and science classes. Most students go to medical school after four years of college, but some colleges, like the University of Michigan, have six- or seven-year programs in which students earn both an undergraduate and a medical degree.

Neurosurgery is a major medical specialty. Neurosurgeons focus on disorders of the nervous system and operate on the brain and spine. Brain abnormalities in children can be caused by disease, injury, and inherited disorders. For example, hydrocephalus is a birth defect in which fluid buildup causes the skull to enlarge. Patients with this disorder may require several surgeries.

Brain tumors are abnormal growths on the brain. A tumor can destroy cells around it and exert pressure that damages other parts of the brain.

Epilepsy is a disease that causes seizures. It is caused by abnormal impulses in cells in one section of the brain. Neurosurgery may be necessary to find the cause of seizures.

1. INTRODUCE

About Alexa Canady: Dr. Alexa Canady comes from a family of high achievers. Her father was a dentist, her mother was the first African-American member of the local school board, and her three brothers are lawyers. Dr. Canady began majoring in math in college but found she was not dedicated to the subject. She entered the University of Michigan's Medical School and excelled, graduating cum laude.

About Canady's Work: Dr. Canady was appointed director of neurosurgery at Children's Hospital in Detroit in 1987. There she operated

on children with brain tumors and with abnormalities that caused seizures. Today she mentors young people. "If you want to be something, you have to perceive that something is possible." However, she never paints too rosy a picture of achieving success. "Everybody fails at some time or other. But no one talks about it." Dr. Canady retired to Pensacola, Florida, with her husband in 2001.

2. DISCUSS

Use the copy master to prompt discussion. Ask students to discuss reasons they might enjoy knowing Dr. Canady or having her as a doctor.

Answers to Explain It:
1. Answers will vary. Possible answer: Yes; it could make a child relaxed, less fearful, and more at home.

2. Answers will vary. Possible answer: A patient needs to communicate with his or her doctor about symptoms and needs to feel free to ask questions and make comments about treatment. Feeling in awe of a doctor would not encourage free communication.

3. EXTEND

After students have completed the Explore It, ask them to share what they have learned. Students should mention wearing a helmet when riding bikes, skating, skateboarding, and playing contact sports; wearing a seat belt in cars; and being careful on playground equipment.

Career Focus: Why is a career in neurosurgery so important? Would you rather work with children or adults?

Annie **Easley**

AT A GLANCE

born 1933
Birmingham, Alabama

education
B.S., Cleveland State University

occupation
Computer scientist

Many people dream of becoming an astronaut for NASA. But not all NASA workers are astronauts. For example, Annie Easley worked at NASA for over 30 years. She did not go into space. Instead, she used her math and computer skills for research. Her work helped send up rockets.

○ **EASLEY** used computers like this.

Easley's work at NASA was based on teamwork. "When you needed to get it done, you all jumped in as a team." She also values communication skills. "You may be the best scientist in the world. But if you can't communicate it to someone else, it does you no good."

? **explain it**

1. Why might a young person dream of becoming an astronaut rather than a computer scientist?

2. Why is communication important for NASA scientists?

! **explore it**

Find out more about NASA (National Aeronautics and Space Administration). When was it started? What are its locations? What are some jobs at NASA?

Annie **Easley**
TEACHING NOTES

Purpose: To introduce students to a computer scientist who worked on a wide variety of projects in her long career at NASA.

Science Background: Computers began with simple punched card systems. Herman Hollerith invented one in 1888 that was used successfully to tabulate the 1890 American census. Hollerith's machine had electrically charged nails that formed a circuit when passed through the holes of punch cards. In 1896, Hollerith founded the Tabulating Machine Company, which became IBM in 1924. Research of electronics led to more powerful computers. The first fully electronic digital computer, called ENIAC, weighed more than 30 tons and took up more than 1,500 square feet of floor space. The machine provided rapid results, but programming it took a long time.

In 1951 computers called EDVAC and UNIVAC were developed that stored more information. Soon transistors were used to replace the vacuum tubes in computers, leading to even greater speed and capacity. Computers became smaller, cheaper, and more readily available in the 1970s with the introduction of silicon chips into computer technology.

I. INTRODUCE

About Annie Easley: Annie Easley started working at NASA in 1955. She took several specialized courses offered there during her employment. In 1977, she obtained her bachelor of science degree in mathematics from Cleveland State University.

About Easley's Work: Easley began working for National Advisory Committee for Aeronautics (NACA), the predecessor to NASA, in Cleveland, Ohio, in 1955. She continued to work with NASA at the Lewis Research Center. As a mathematician, she supported engineers involved in space research.

When she began, calculations were done with paper and pencil, slide rules, and calculators. She later used punch cards submitted to a main frame computer before reaching today's computer age. Easley develops and implements computer codes used in solar, wind and other energy projects. She says, "In my lifetime, to have seen where we were and where we are . . . we can just do things instantly now with the equipment that's available."

2. DISCUSS

Use the copy master to prompt discussion. Discuss the fact that sending astronauts into space requires extensive behind-the-scenes work by hundreds of people. Point out that although astronauts become famous, unknown workers like Annie Easley make the feats of the astronauts possible.

Answers to Explain It:

1. Answers will vary. Possible answer: The feats of astronauts are widely publicized; young people are impressed by the excitement and glory of going into space. Behind-the-scene workers like computer scientists do not get much attention.

2. Answers will vary. Possible answer: Many different kinds of scientists—engineers, physicists, astronomers, and computer scientists—must work together to accomplish goals when studying space. They must be able to communicate their ideas and research to one another in order to achieve a goal together.

3. EXTEND

After students have completed the Explore It, ask them to share what they have learned. NASA was established in 1958 as an independent agency. It has 10 major facilities, including Kennedy Space Center in Florida and Johnson Space Center in Houston. Careers include scientists, engineers, and computer scientists.

Career Focus: Computer-related occupations continue to employ many people. What kind of computer skills are needed by people in other jobs?

Philip Emeagwali

AT A GLANCE

born 1954
Akure, Nigeria

education
M.A., University of Maryland
Ph.D., University of Michigan

occupation
Computer scientist

"Trying to study computer science without a good knowledge of math is like trying to play soccer without learning to run," says Philip Emeagwali. He should know.

Imagine linking the power of 65,000 computers. Experts said it wouldn't work. But Dr. Emeagwali did it, and he won the top computer science prize.

EMEAGWALI helped start the World Wide Web.

Emeagwali had to quit school after the seventh grade. His native country, Nigeria, was at war. But his father had him solve math problems each day: 100 an hour!

Dr. Emeagwali says, "I was not born a genius. It was nurtured in me by my father."

? explain it

1. Why would winning the top computer prize be important?
2. How do you think the hardships in Emeagwali's early life influenced him?
3. Did Emeagwali's father make his son into a genius? Explain.

! explore it

Emeagwali is known as one of the founding fathers of the Internet. Find three or more facts about the history of the Internet. For example, you can find out what the Internet was first called, and when it was first used.

Philip Emeagwali

TEACHING NOTES

Purpose: To introduce students to one of the world's top computer scientists.

Science Background: A supercomputer is a computer that has greater processing capacity (usually based on its speed of calculation) than other computers of its time. The term has been used since 1920, and so the definition of a supercomputer is constantly changing. Today a supercomputer is often a cluster of machines in which many high-performance computers work in parallel. Emeagwali says that his use of the Connection Machine required an "absolute understanding of how all 65,536 processors are interconnected. The processing nodes are configured as a cube in a 12-dimensional universe, although we only use it to solve problems arising from our three-dimensional universe." His programming of the machine required 1,057 pages of equations, algorithms, and programming techniques. In 1990 Emeagwali proved that the speed of supercomputers can be increased one million times. He says that the greatly increased processing speed of supercomputers can ensure that computers don't need downtime and makes them safe from hackers. Supercomputers can also keep the response time constant for high-volume websites such as Internet search engines.

I. INTRODUCE

About Philip Emeagwali: During the Nigerian civil war (1967-1970) Emeagwali's family was homeless and lived in various refugee camps. Emeagwali left Nigeria and went to London as a young man. Studying on his own, he earned a general education certificate from the University of London. He came to the U.S. in 1974 after winning a full scholarship to Oregon State University. He earned two M.A.s from George Washington University and one from the University of Maryland.

About Emeagwali's work: Emeagwali is known as one of the founding fathers of the Internet for his work on harnessing the power of thousands of computers. The government developed the Connection Machine and then considered it useless. Emeagwali showed how it could work. His computer studies have shown how oil flows underground so it can be obtained more efficiently, and have made weather forecasts more accurate. Emeagwali has also used computers to help predict the spread of AIDS. He says, "My focus is not on solving nature's deeper mysteries. It is on using nature's deeper mysteries to solve important societal problems."

2. DISCUSS

Use the copy master to prompt discussion. Have students explain Emeagwali's comment comparing computer science and soccer. Ask them to describe the importance Emeagwali puts on math study.

Answers to Explain It:
1. Answers will vary. Possible answer: The prize probably gave Emeagwali confidence, showed others that he was a good programmer, and perhaps helped him financially.

2. Answers will vary. Possible answer: Emeagwali's hardships probably taught him the importance of hard work in achieving goals; it may also have given him the urge to help others.

3. Answers will vary. Possible answer: Emeagwali's father probably could not make his son into a genius, but his exercise probably helped Philip think faster and gave him self-discipline.

3. EXTEND

After students have completed the Explore It, ask them to share what they have learned. The Internet was first commissioned by the U.S. Air Force. Paul Baran had the idea of relaying data from one computer to the next in 1962. In 1969, ARPANET was first used by UCLA, Stanford, UC Santa Barbara, and the University of Utah.

Career Focus: Would you like to be a computer scientist like Emeagwali? Why or why not?

Evelyn Granville

AT A GLANCE

born 1924
Washington, D.C.

education
A.B., Smith College
Ph.D., Yale University

occupation
Mathematician

Which is harder—sending rockets into space or teaching math to children? You could ask Evelyn Granville. She has done both. And she found both jobs interesting and challenging.

In college, Dr. Granville almost majored in astronomy. But she did not know that the American space program would soon take off. So she became a mathematician. At IBM, she worked on Project Mercury. This was the first U.S. program to send people into space. She said this was the most interesting job of her life. Later Dr. Granville taught math.

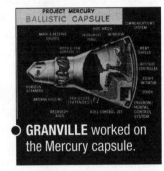

GRANVILLE worked on the Mercury capsule.

? explain it

1. How is teaching math as challenging as rocket science?

2. Suppose Dr. Granville had known the space program would soon be launched. Why would this have changed her plan?

! explore it

Find out about Project Vanguard and Project Mercury.
When did U.S. scientists work on them? What were their goals?

Evelyn Granville
TEACHING NOTES

Purpose: To introduce students to a mathematician who worked on computations for the space program and later became a math educator.

Science Background: In 1961, the Soviet Union became the first country to send a manned spacecraft into orbit in space. The first American astronaut to orbit Earth in a spacecraft was John Glenn, in 1962. This was part of the Mercury project.

One of the first steps in sending a spacecraft into space is selecting an orbit. Early crewed spacecraft usually orbited less than 200 miles above Earth. This helped them avoid the Van Allen belts, zones of electrically charged particles surrounding Earth.

A spacecraft may have a circular orbit, always staying the same distance from Earth, or an elliptical orbit, changing its distance from Earth continually. In a circular orbit, the spacecraft's speed stays constant; in an elliptical orbit, it changes continually.

A spacecraft is launched by a rocket that sends it in a particular direction based on the inclination chosen for it. Inclination is the angle between the plane of the spacecraft's orbit and the equator. To achieve an orbit, the speed of the spacecraft must be balanced with gravity. If the spacecraft moves too slowly, gravity will pull it back to Earth. If it moves too quickly, gravity will not be able to keep it in orbit and it will drift into space. The speed at which the spacecraft is launched is also based on orbit height. If a spacecraft has a high orbit, it requires less speed to stay in orbit. A spacecraft with a low orbit requires more speed.

1. INTRODUCE

About Evelyn Granville: Going to school in Washington, D.C., Evelyn Granville got much encouragement and an excellent education. She graduated with high honors from Smith College. She attended Yale University and became one of the first African-American women to earn a Ph.D. in mathematics.

About Granville's Work: From 1956 to 1960, Dr. Granville worked at IBM on orbit computation and computer procedures for two space programs: Project Vanguard and Project Mercury. She said, "That was exciting . . . to be a part of the space programs—a very small part—at the beginning of U.S. involvement." She later taught at California State University–Los Angeles and Texas College. "Over the years I have learned that one of the best ways to become proficient in a subject is to teach the subject." Asked to name her major accomplishments, Dr. Granville said, "First of all, showing that women can do mathematics."

2. DISCUSS

Use the copy master to prompt discussion. Discuss math in space travel. For example, a spacecraft orbiting 200 miles above Earth must travel at a speed higher than 17,000 mph.

Answers to Explain It:

1. Answers will vary. Possible answer: Both require a thorough knowledge of math. Both require focus and problem solving.

2. Answers will vary. Possible answer: Dr. Granville probably feared she would not find work in astronomy while there were many careers available in math. If she had known that NASA would soon be needing many different kinds of mathematicians and scientists for its space program, she would have realized that there would be many career opportunities in astronomy as well.

3. EXTEND

After students have completed the Explore It, ask them to share what they have learned. Project Vanguard, 1955, was the first official space satellite program. Project Mercury, 1958–1963, sought to send a man into space.

Career Focus: What kinds of careers require a solid mathematics education?

Shirley Ann Jackson

AT A GLANCE

born 1946
Washington, D.C.

education
B.S., Ph.D., Massachusetts Institute of Technology (MIT)

occupation
Physicist

Shirley Ann Jackson is "a woman of astonishing firsts." She was the first African-American woman to earn a Ph.D. from MIT. She was the first to lead a national research university. She also led a government group that was in charge of using nuclear energy wisely.

Scientists get nuclear energy by splitting atoms or joining them together. Dr. Jackson's work with atoms is key in making computers and lasers.

Dr. Jackson raised bees as a child. She also built go-carts. Now she uses physics to help people understand the world around them.

JACKSON'S research is used to develop lasers.

? explain it

1. What do Jackson's childhood interests show about her?
2. What are some ways that science helps people understand the world better?

! explore it

Use the Internet to find out about the three parts of an atom. What kind of electric charge does each part have?

Shirley Ann Jackson
TEACHING NOTES

Purpose: To introduce students to a scientist who is a respected physicist and academic leader in the United States today.

Science Background: Physics is the study of matter and energy. A theoretical physicist like Shirley Ann Jackson develops laws and theories. She has focused on condensed matter physics, the study of the electronic structure of solids. This research is key in understanding superconductors, which have no electrical resistance, and semiconductors, which are used in modern electronic devices. Dr. Jackson's research has centered on properties of electrons in two-dimensional systems, namely their conductivity and interactions with light. This research is used in developing integrated circuits and semiconductor lasers.

A semiconductor is a material such as silicon that conducts electricity better than insulators such as glass but not as well as conductors like copper. In a conductor such as copper wire, electrons move freely from atom to atom, forming an electric current. An n-type semiconductor is made when impurities such as arsenic are present in an insulator. The impurities allow a few free electrons to form an electric current. A p-type semiconductor has impurities like aluminum. They take electrons away from some atoms, making holes that flow from atom to atom, forming an electric current.

1. INTRODUCE

About Shirley Ann Jackson: Shirley Ann Jackson's father encouraged her to "aim for the stars." She took that advice and graduated valedictorian from the segregated Roosevelt High School in Washington, D.C. Dr. Jackson was the first African-American woman to receive a Ph.D. from MIT. She was inducted into the National Women's Hall of Fame in 1998.

About Jackson's Work: Dr. Jackson is a prominent scientist and a leader in many different realms. In 1995 President Clinton appointed Jackson head of the U.S. Nuclear Regulatory Commission, the first African American to serve in this role. She served until 1999 when she became president of Rensselaer Polytechnic Institute. As a scientist and academic leader today she notes that "the boundaries of specialization are blurring…many of the most exciting frontiers of scientific research today represent a convergence of disciplinary forces, creating new discoveries and often new sciences."

2. DISCUSS

Use the copy master to prompt discussion. Discuss nuclear energy, explaining that it has been used for weapons but also as an energy source.

Answers to Explain It:
1. Answers will vary. Possible answer: Dr. Jackson's interests show that she was curious and involved, and that she was interested in both the natural world and in applying science to make interesting things.

2. Answers will vary. Possible answer: Different kinds of science lead to understanding of various parts of our world. For example, biology helps us understand plants and animals. Astronomy helps us understand the stars, planets, and other parts of the universe. Physics helps us understand what matter is made of and how we can create and use electricity and other types of energy.

3. EXTEND

After students have completed the Explore It, ask them to share what they have learned. Discuss the three parts of the atom: the proton, with a positive electric charge; the electron, with a negative electric charge; and the neutron, with no electric charge. Explain that atoms are the building blocks of matter. Research of the parts of the atom has led to important developments in electronics and nuclear science.

Career Focus: If you became a physicist, what kind of work would you like to do?

Delano Meriwether

AT A GLANCE

● **born** 1944
education
M.D., Duke University
Medical School
occupation
Physician

What does a track star know about flu shots? Ask Dr. Delano Meriwether! In 1963, he was the first African American to attend Duke University Medical School. He became champion in the 100-yard dash in 1971.

In 1976, he worked for the U.S. assistant secretary for health. Swine flu, a new strain of flu, appeared.

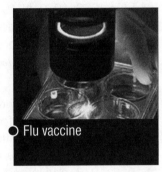

○ Flu vaccine

Doctors feared it could kill thousands. President Ford began a flu shot program. Dr. Meriwether helped run the program.

Later he worked as a doctor in Africa, the most rewarding experience of his life.

? **explain it**

1. How might Dr. Meriwether's experience as a runner help him in his medical career?

2. Why do you think working in Africa was so rewarding?

! **explore it**

What are the symptoms of flu? Why can it be so dangerous? Find out more about influenza in the library or on the Internet.

Delano Meriwether
TEACHING NOTES

Purpose: To introduce students to a physician who was a public health official involved in the swine flu inoculation program of 1976.

Science Background: Influenza, commonly known as the flu, is an infectious disease caused by a virus. The disease affects the respiratory system and causes fever, chills, headache, and body aches that last about a week. Each flu outbreak is caused by a different form of the virus, or strain, than the previous one. For example, the Hong Kong flu spread around the world in 1968–1969. In 1918–1919, a strain of flu killed about 20 million people around the world. Vaccines can limit the spread of flu, but scientists must create a new vaccine for each new flu strain.

In August 1976, an Army recruit in New Jersey fell ill and died. Doctors found he had a new strain of flu called swine flu because it usually affected pigs. The virus was similar to the one of 1918–1919, and experts feared it could kill many people. President Ford called for mass inoculations. Two hundred million doses of swine flu vaccine were prepared. However, only about 24 percent of Americans received inoculations because the expected epidemic did not occur. Some people criticized the government for overreacting. But most believed the government showed it could inoculate many people quickly.

1. INTRODUCE

About Delano Meriwether: Delano Meriwether was a hematologist (doctor of blood diseases) at Baltimore Cancer Research Center in 1970. He was also a runner. He ran six major races, winning two, in the next year. At the National Outdoor Championships, he won the 100-yard dash. He said, "My body is exercising and my mind is relaxing. I think these are the essentials of the sport."

About Meriwether's Work: Meriwether went to work at Harvard Medical School before moving to South Africa in 1983. There he worked as one of only six doctors for about half a million people. He returned to the medical profession in the United States in 1990.

2. DISCUSS

Use the copy master to prompt discussion. Have students discuss common traits Meriwether showed in his running career, in his role in the swine flu situation, and in his work in Africa.

Answers to Explain It:

1. Answers will vary. Possible answer: The endurance and training required in running might be similar to those required of a doctor. They might train him to do the hard work necessary to be a good doctor.

2. Answers will vary. Possible answer: Africans are not as fortunate as Americans in having excellent, easy-to-obtain health care. Meriwether was probably able to help many people who would not otherwise have had medical care.

3. EXTEND

After students have completed the Explore It, ask them to share what they have learned. Discuss the fact that people may mistakenly call a bad cold "flu." Point out that influenza is more serious than a cold, and that it can lead to serious complications such as pneumonia in old people and very young people.

Career Focus: Explain that scientists who study the process by which the body produces disease-fighting cells and antibodies are called immunologists. What role do immunologists play in helping the public stay healthy?

Ruth Ella Moore

AT A GLANCE

lived 1903–1994
education
B.S., M.A., Ph.D., Ohio State
University
occupation
Bacteriologist

About 100 years ago, many people were dying of a disease called tuberculosis, or TB. Ruth Ella Moore was an early researcher of the TB bacteria.

Dr. Moore was the first African American to earn a Ph.D. in bacteriology. Bacteriologists study bacteria. These are tiny organisms that cause disease. Dr. Moore studied bacteria like TB, and salmonella which is found in some foods.

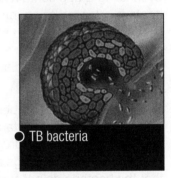
○ TB bacteria

Dr. Moore was also the first woman to head a medical school department.

? **explain it**

1. Why do you think Ruth Ella Moore chose to research the TB bacteria?
2. Why is it important to find the cause of a disease?

! **explore it**

Find out two ways a person can be infected with the bacteria that cause TB.

Ruth Ella Moore
TEACHING NOTES

Purpose: To introduce students to a scientist who studied microbes that cause serious illnesses.

Science Background: Tuberculosis was once one of the world's most common causes of death. In 1882 Robert Koch, a German physicist, found that tuberculosis was caused by rod-shaped bacteria called tubercle bacilli. Bacteria are among the smallest single-celled organisms. Tuberculosis is most often contracted by inhaling drops containing the tuberculosis bacteria released as an infected person sneezes or coughs. People may have a primary infection with no symptoms and a secondary infection that occurs much later. The disease may get progressively worse for years.

The first treatment for tuberculosis, in the 1800s, was a rest cure at a sanitarium. In the 1940s researchers developed drugs to treat the disease. Today several drugs are used at once so that the bacteria do not become resistant to one drug. Skin tests can indicate the presence of tubercle bacilli in the body. Tuberculosis had not been a major problem in the United States and other developed countries for the past 50 years. However, the disease became more common between 1985 and 1992 due to complacency and the emergence of HIV/AIDS, which lowers a person's immunity. Today the disease is once again rare in the United States, but it is still a problem in developing nations.

1. INTRODUCE

About Ruth Ella Moore: Ask students: What traits do you think Moore had in order to be the first African American to earn a Ph.D. in bacteriology and the first woman to head a medical school department?

About Moore's Work: Ruth Ella Moore was the first woman chair of the Department of Bacteriology at Howard University Medical College. Moore was a professor at Howard until her retirement. The focus of her research was on blood grouping (e.g., defining blood types such as A, B, or O) and enteriobacteriaceae, a large family of bacteria.

2. DISCUSS

Use the copy master to prompt discussion. Discuss the years of hard work that go into studying a science such as bacteriology. Help students realize that this work would be even more difficult for a person such as Ruth Ella Moore, who was a pioneer both as an African American and as a woman.

Answers to Explain It:

1. Answers will vary. Students may say that Moore was probably familiar with the suffering caused by tuberculosis and hoped to help stop this suffering by finding a cure for the disease.

2. Answers will vary. Possible answer: It is necessary to find the cause of a disease in order to understand how to prevent it or to develop ways to cure it.

3. EXTEND

After students have completed the Explore It, ask them to share what they have learned. Discuss the fact that most people get tuberculosis from inhaling droplets of the tuberculosis bacteria from infected individuals. People can also get the disease from consuming infected food or milk from an animal with the disease. Pasteurizing the milk destroys this bacteria from infected cows.

Career Focus: Medical microbiologists, like Ruth Ella Moore, study microbes such as viruses and bacteria. Why is it important to study bacteria and viruses? In what career fields can mircobiologists find work?

David Robinson

AT A GLANCE

born 1965
Key West, Florida

education
U.S. Naval Academy

occupation
Former professional
basketball player

Can you be one of the greatest NBA players in history and enjoy math? David Robinson would say yes. Robinson got his math degree at the U.S. Naval Academy.

Robinson was a star player at the Academy. When he graduated, the San Antonio Spurs drafted him. But he kept his promise to the Navy. He served for two years.

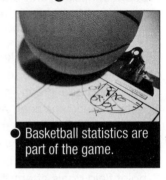

Basketball statistics are part of the game.

Robinson was NBA's Most Valuable Player in 1995. Math is used to determine how good a player is at shooting free throws and how often any shot is missed.

The NBA named a community service award for him.

? explain it

1. In what ways would a knowledge of math help a basketball player?
2. Why did Robinson choose go to college and serve in the Navy before becoming an NBA player?

! explore it

Statistics are facts that are expressed with numbers. Statistics are often used in sports to describe players and games. Find out about three statistics that are often used in basketball. Explain each statistic.

David **Robinson**
TEACHING NOTES

Purpose: To introduce students to a professional basketball player who got a math degree before playing in the NBA.

Mathematics Background: Basketball, like many sports, revolves around statistics. Coaches, players, and fans assess games and players by studying statistics about field goals, rebounds, and free throws. For example, David Robinson is one of only four players to achieve a quadruple-double: double digits in points scored, rebounds, assists, and blocks in one game (34 points, 10 rebounds, 10 assists, 10 blocks against the Detroit Pistons, 1994).

Like the game itself, basketball statistics are constantly evolving. For example, instituting the 3-point field goal greatly changed the game. Sports statisticians such as Thomas P. Ryan suggested that statistics needed to be devised to reflect these changes. Ryan said that fans "need to know the relationship between the number of 3-point field goals attempted and the number of 2-point field goals attempted." He suggested the following formula for the composite field-goal percentage:

$$C = (a + 1.5b)/N$$

a = number of 2-point field goals made

b = number of 3-point field goals made

N = total number of field-goal attempts

Basketball statistics can enliven math for students. For example, younger students can order players by height, calculate score differentials, and figure scoring percentages.

I. INTRODUCE

About David Robinson: After achieving high SAT scores, Robinson was admitted to the U.S. Naval Academy. He played on the U.S. Olympic basketball teams in 1988, 1992, and 1996. Robinson was drafted by the San Antonio Spurs out of college but put his pro career on hold to serve in the Navy for two years.

About Robinson's Work: As a basketball player, Robinson was known for "Mr. Robinson's neighborhood," a section of seats in the San Antonio Spurs' stadium reserved for underprivileged families. Since retiring from pro basketball in 2003, Robinson has continued the philanthropy. He and his wife contributed $11 million to The Carver Academy (named after George Washington Carver) in San Antonio, which educates inner-city elementary school children. Robinson said, "My career as a sports figure has been exciting, but its main purpose is to provide a platform for me to impact people's lives in a positive way."

2. DISCUSS

Use the copy master to prompt discussion. Have students discuss the character traits shown by Robinson on and off the basketball court.

Answers to Explain It:
1. Answers will vary. Possible answers: Understanding statistics can show players their strengths and weaknesses as well as those of others; the focus used in math could transfer to the basketball court.

2. Answers will vary. Possible answer: Getting an education and serving his country were higher priorities than being a pro athlete.

3. EXTEND

After students have completed the Explore It, ask them to share what they have learned. Statistics include field goal, free throw, and rebound percentages for both individuals and teams. Help students understand how a field goal percentage, for example, is calculated. If a player attempts 12 field goals and makes 6, his percentage is 6 ÷ 12, or 50%.

Career Focus: Why would earning a college degree in math be wiser than focusing on playing basketball with hopes of becoming a pro player?

Earl **Shaw**

AT A GLANCE

born 1937
Clarksdale, Mississippi

education
B.S., University of Illinois
M.A., Dartmouth College
Ph.D., University of California

occupation
Physicist, Professor

What could you do with a "spin-flip laser"? No, it's not a backyard toy. It is an important scientific tool. A laser makes a strong beam of light. One of its uses is to study living cells.

Dr. Earl Shaw invented the spin-flip laser. The spin-flip laser allows scientists to study cells without damaging them.

○ **SHAW** at work

Dr. Shaw also is a professor of physics. He enjoys helping students like Charles Kopec. Dr. Shaw helped Kopec get a telescope for the college.

"Dr. Shaw encourages me to take things to the next level and to explore my ideas," Kopec said.

? explain it

1. How does Dr. Earl Shaw's story show that scientists work together?
2. Do you think Dr. Shaw is a good teacher? Explain why or why not.
3. What does Charles Kopec's statement tell you about Dr. Shaw?

! explore it

Find out one or more uses for lasers in each of the following fields: medicine, industry, military, and scientific research.

Earl Shaw
TEACHING NOTES

Purpose: To introduce students to a physicist who invented an important new type of laser.

Science Background: Electromagnetic waves are related patterns of electric and magnetic force that travel through space at the speed of light. A wavelength varies in strength, having crests and troughs. The distance from one crest to another is a wavelength. The electromagnetic spectrum is made up of bands of different wavelengths. The spectrum ranges from gamma rays, with the shortest wavelength, to radio waves, with the longest wavelength. The longest waves are easiest to produce. The use of radio waves for communication began around 1900. Lasers make use of some of the shorter wavelengths.

The frequency of a wavelength is the number of vibrations it has in a second. Lasers have a narrow frequency range on the electromagnetic spectrum, whereas light from other sources has many frequencies. Lasers amplify light, making a thin, intense beam. Unlike light from other sources such as a lightbulb or a flashlight, the light from a laser travels in only one direction. A laser focuses light like a flashlight. However, the light is usually all the same frequency. The spin-flip tunable laser works like a light with a dimmer switch. It allows the operator to adjust the laser so that it has a stronger beam and a different color. (The color is not visible to the unaided eye.) This tool allows scientists and others to have greater control in the laser strength they use.

I. INTRODUCE

About Earl Shaw: Dr. Earl Shaw attended a three-room school in Mississippi until he moved to Chicago at age 11. Although he felt that he did not receive a good general education at his high school, he became fascinated by physics. He earned a B.S. in physics at the University of Illinois and went on to earn an M.A. at Dartmouth and a Ph.D. at the University of California, Berkeley.

About Shaw's Work: While working as a research scientist at Bell Laboratories, Dr. Shaw helped invent the spin-flip tunable laser. He is a professor at Rutgers-Newark University and served as the chair of its Physics Department.

2. DISCUSS

Use the copy master to prompt discussion. Have students discuss traits, such as creativity, cooperation, involvement, and generosity, shown by Earl Shaw in his various roles.

Answers to Explain It:

1. Answers will vary. Possible answer: Shaw's work as a physicist to improve lasers helped scientists such as biologists; he also has helped students learn more about another science, astronomy.

2. Answers will vary. Possible answer: Yes, because a former student described him as encouraging.

3. Answers will vary. Possible answer: It is a first-hand account of Shaw's effectiveness in his job.

3. EXTEND

After students have completed the Explore It, ask them to share what they have learned. Uses for lasers include the following: medicine—removing diseased body tissue, sealing off blood vessels cut during surgery, correcting the eye condition retinal detachment; industry—cutting blades, drilling eyes in surgical needles, guiding heavy machinery, creating heat to melt hard materials; military—guiding bombs and missiles, locating enemy targets; science—creating plasmas, studying cells; entertainment—DVD players and CD players.

Career Focus: What type of fields would a laser technician work in?

James Tilmon

AT A GLANCE

born 1934
Guthrie, Oklahoma

education
B.A., Lincoln University

occupation
Meteorologist, Pilot

Would you like to be a pilot or a weather forecaster? How about a reporter or a musician? If you were Jim Tilmon, you would not choose just one career. You would be all these things!

In 1965, Tilmon became a pilot for American Airlines. He was the third African American to fly for an airline.

Pilots need to know about the weather.

While Tilmon was a pilot, he became a TV weather forecaster. He was a weatherman on Chicago TV for 18 years. When he retired, he did not stop working. Tilmon often appears on TV to report on weather and air travel.

? explain it

1. Why do airplane pilots need to know about weather?
2. What traits do you think a TV weather forecaster should have?

! explore it

What is a jet stream? Check the Internet to find out why airplane pilots need to know about jet streams.

James Tilmon
TEACHING NOTES

Purpose: To introduce students to a meteorologist who has had careers as a TV weather forecaster and an airline pilot.

Science Background: Meteorology is the study of Earth's atmosphere and the weather conditions produced by atmospheric changes. Meteorologists study the main factors in weather: wind, temperature, precipitation, and air pressure.

Meteorology is an important part of ground study for airplane pilots. Important meteorological topics for pilots include the following:

• The levels of the atmosphere: troposphere, stratosphere, mesosphere, and thermosphere. Oxygen decreases in the atmosphere above 10,000–12,000 feet.

• Types of clouds: Different types of clouds bring different types of weather, such as rain and thunderstorms, and various degrees of turbulence to an aircraft.

• Winds: Earth has prevailing winds, or the winds that usually occur over large areas. Prevailing winds and local winds can cause aircraft turbulence, slow planes down, or help planes go faster. Winds behave differently over flat land, mountains, and oceans. For example, pilots need to know about potential dangers of winds over mountains.

1. INTRODUCE

About James Tilmon: James Tilmon earned a bachelor of arts degree in music but was interested in engineering as well. He worked for the U.S. Army Corps of Engineers before training as a pilot.

About Tilmon's Work: Tilmon was a pilot for American Airlines from 1965 until 1994. He was awarded the Captain's Chair Award by American Airlines and named an honorary pilot by United

Airlines. Tilmon received national notoriety for his role in developing *Our People,* a television program for African Americans. It was revolutionary when it premiered in 1968. He then became a weather forecaster on Chicago TV from 1974 to 1988 and from 1990 to 1994 and won an Emmy in 1994. After his retirement, Tilmon still reported on TV about aviation and weather-related topics. He returned to weather forecasting in Chicago in 2002.

2. DISCUSS

Use the copy master to prompt discussion. Have students explore the character traits that likely made Tilmon successful in so many different fields.

Answers to Explain It:
1. Answers will vary. Possible answer: Pilots need to know about weather because many aspects of weather affect the way an airplane flies. For example, a thunderstorm may be dangerous to a plane; winds may slow a plane down or speed it up; oxygen may be needed at certain altitudes. Understanding weather helps a pilot fly safely.

2. Answers will vary. Possible answer: TV weather forecasters need to know about the causes and effects of weather. They need to be personable and speak clearly. They need to be able to explain weather clearly so that everyone can understand them.

3. EXTEND

After students have completed the Explore It, ask them to share what they have learned. A jet stream is a band of fast-moving air currents at high altitudes. It can affect a plane's speed and cause turbulence.

Career Focus: Meteorologists work mainly as TV weather forecasters. But do you know what other fields they work in?

Warren M. Washington

AT A GLANCE

- **born** 1936
  Portland, Oregon
- **education**
  B.A., Oregon State University
  M.A., Ph.D., Pennsylvania State
- **occupation**
  Meteorologist

What do icebergs have to do with global warming? Ask Dr. Warren Washington! He heads a group at the National Center for Atmospheric Research in Colorado. They study climate change. Dr. Washington traveled to the North and South Poles. "Polar regions are key to future climate change," Dr. Washington said.

Global warming affects life in polar regions.

Global warming can affect all forms of life. Dr. Washington's team found that the oceans soak up much heat. This could slow down global warming. But how much? Their studies may find the answer.

❓ explain it

1. Why do people need to learn about future weather? Think about reasons related to both work and fun.
2. Why is it important to study global warming?

❗ explore it

Find out more about global warming. Find out how people may be causing it. Find out what could happen as a result of global warming.

Warren M. Washington

TEACHING NOTES

Purpose: To introduce students to a scientist who is in the forefront of research on future climate change.

Science Background: Climate is the average of weather conditions of a given area over a long period. The two conditions that determine climate are temperature and precipitation. Each is affected by factors such as winds, ocean currents, landforms, latitude, and altitude.

Earth's atmosphere contains "greenhouse" gases such as carbon dioxide that keep Earth warm. Since the Industrial Revolution, people have increased the amount of greenhouse gases in the atmosphere. Industry, automobiles, and heating use fossil fuels, which release these gases.

Global warming is the increase of Earth's average temperature. Global warming can change world rainfall totals and increase the sea level, possibly affecting all forms of life. Earth's average temperature has increased by 1° in the past 100 years. The four warmest years of the 20th century were all in the 1990s. Is this warming a natural phenomenon or have people contributed to it through the burning of fossil fuels? This is one of the questions that researchers such as Warren Washington seek to answer. They study the atmosphere, the oceans, sea ice, and land formations to develop computer climate models.

I. INTRODUCE

About Warren Washington: Explain that Dr. Washington's interest in becoming a scientist began when he asked a question in a high school chemistry class. Instead of answering him directly, the teacher discussed ways to find the answer. Dr. Washington soon began to enjoy scientific research. On his trip to Antarctica to examine the polar ice cap, Washington also enjoyed photographing the penguins that lived there.

About Washington's Work: Dr. Washington worked as a research assistant at Pennsylvania State University and taught as an adjunct professor at the University of Michigan. He is currently the director of the Climate Change Research Station of the Climate and Global Dynamics Division for the National Center for Atmospheric Research (NCAR) in Boulder, Colorado, where he tracks violent weather and studies global warming.

2. DISCUSS

Use the copy master to prompt discussion. Point out the difference between weather—the atmospheric conditions at a given place and time—and climate—the average weather conditions of an area over a long period. Ask which parts of Washington's job students would enjoy. Point out that travel to interesting places and studying computer models are both key parts of Washington's job. Ask students why researching Earth's future climate would be interesting and challenging.

Answers to Explain It:

1. Answers will vary. Possible answer: People need to know about the weather so that they know how warmly to dress for work and school, can plan outdoor activities such as sports and picnics, and can plan travel. Weather forecasts are especially important for farmers, construction workers, and landscapers.

2. Answers will vary. Possible answer: It is important to study global warming so that people can learn whether they are contributing to it and, if so, how to reverse it. They also need to find out its possible effects and whether there is any way to avoid them.

3. EXTEND

After students have completed the Explore It, ask them to share what they have learned. Discuss the definition and possible causes of global warming.

Career Focus: Computers and technology have made studying both weather and climate much easier for meteorologists. Discuss computers and radar, two tools meteorologists must learn to use.

Name	Date

David Blackwell

AT A GLANCE

born 1919
Centralia, Illinois

education
B.S., M.S., Ph.D.,
University of Illinois

occupation
Mathematician

Dr. David Blackwell grew up in a small town in Illinois. He was lucky to have supportive parents, who taught him about the value of hard work. They encouraged him to get a good education.

Blackwell studied hard in school. He liked geometry. When he took a course in analysis, he realized that "serious mathematics was for me." He recalled, "It became clear that it was not simply a few things that I liked. The whole subject was just beautiful."

BLACKWELL enjoys solving math problems.

When Blackwell entered the University of Illinois, he was only 16. At the age of 22, he earned his doctorate degree. He was the youngest African American to earn a Ph.D. in mathematics.

Dr. Blackwell built a successful career in mathematics. He taught math at respected universities and headed several math departments. In addition, he wrote books and papers on statistics and probability.

? **explain it**

1. What made Dr. Blackwell realize that math was "beautiful"?
2. How did Blackwell's family help him become successful?

! **explore it**

In 1965 Dr. Blackwell became the first African American named to the National Academy of Sciences. Find out about the Academy. Who are some of the other members? What must a person do to gain entrance into the Academy?

David Blackwell

TEACHING NOTES

Purpose: To introduce students to a successful mathematician who earned his Ph.D. in mathematics at the age of 22.

Mathematics Background:
Although Dr. Blackwell contributed to many areas in mathematics, he devoted much of his time to statistics and probability theory. Statistics is an applied mathematics science that deals with the collection, organization, and interpretation of quantitative data. Probability theory deals with whether events or actions occur independently from one another. This includes the law of averages. Statisticians develop models based on probability theory, which is itself a major branch of mathematics.

Statistics and probability theory have widespread applications in science, engineering, and many other areas. For example, models have been developed to improve radio communications. Sunspots and solar flares can disrupt radio and television signals.

Statistics and probability theory are often used in weather forecasting, market research, health studies, opinion polling, and many other areas.

1. INTRODUCE

About David Blackwell: Dr. Blackwell grew up in southern Illinois in a small town. His parents had high expectations for their son. Consequently, he entered college at the age of 16. He received his bachelor's, master's, and doctorate degrees from the University of Illinois in Urbana–Champaign.

About Blackwell's Work: Dr. Blackwell taught at Southern University and Clark College before joining the teaching staff at Howard University in 1943. He served as the chair of the school's Mathematics Department from 1947 to 1954, publishing more than 20 research papers. Dr. Blackwell then accepted an invitation to join the Department of Statistics at the University of California in Berkeley. He remained there for the rest of his career, serving as the department's chair for several years. Through his research, Dr. Blackwell has contributed to various areas in mathematics, including game theory, probability theory, and mathematical statistics. He is the co-author of *Theory of Games and Statistical Decisions*.

2. DISCUSS

Use the copy master to prompt discussion. Help students understand that many successful people can recall one event in their lives that sparked their interest in a given field or subject. For Blackwell, this spark was provided by the analysis course that he took in high school.

Answers to Explain It:
1. Dr. Blackwell realized that math was beautiful after taking a course in analysis.

2. Answers will vary. Possible answer: Dr. Blackwell learned about the value of hard work from his family. His family probably also taught him other values that contributed to his success.

3. EXTEND

After students have completed the Explore It, ask them to share what they have learned. You might like to direct students to the National Academy of Sciences website:

http://nationalacademies.org

Students can locate members by profession as well as by last name. Dr. Blackwell is located in the applied mathematical sciences category.

Career Focus: What are some of your interests and how did you develop them? Write a paragraph describing how you could develop one or several of these interests into a career.

George Carruthers

AT A GLANCE

born 1939
Cincinnati, Ohio

education
B.S., M.S., Ph.D.,
University of Illinois

occupation
Inventor, Astrophysicist

George Carruthers has always been fascinated with space and space travel. When he was 10, he built his own telescope. Not surprisingly, Carruthers studied astronomy in college. In 1964 he earned a Ph.D. and was invited to study rocket astronomy at the Naval Research Laboratory in Washington, D.C.

In 1969 Dr. Carruthers invented a camera designed to take pictures of deep space and Earth's atmosphere. In 1972 astronauts carried his camera to the moon. Because of its clear images, scientists learned new things about Earth's atmosphere. For example, they learned new ways to control air pollution.

A comet captured by Carruthers' camera

NASA has used another camera Dr. Carruthers built on space shuttle flights to capture images of distant stars and comets.

In 2003 Dr. Carruthers was named a member of the National Inventors Hall of Fame.

? explain it

1. Why is it important to study Earth's atmosphere and objects in deep space?
2. Would you consider Dr. Carruthers a pioneer? Why or why not?

! explore it

Dr. Carruthers' cameras have taken many interesting pictures of Earth's atmosphere, comets, and the Milky Way. Find some of these pictures on the Internet. Print and assemble some of your favorites in a scrapbook.

George **Carruthers**
TEACHING NOTES

Purpose: To introduce students to an astrophysicist and inventor who designed the camera that was used on the Apollo 16 moon mission.

Science Background: The 50-pound, gold-plated Far-Ultraviolet Camera/Spectograph—a combination spectrograph and camera with an electron intensifier—was mounted on a tripod by astronauts during the Apollo 16 mission in 1972. It took more than 200 pictures of Earth's atmosphere, the Milky Way Galaxy, and other deep-space objects. The ultraviolet light used in the camera took pictures that would have been impossible to take using visible light. Photographs from the Apollo 16 mission show dense concentrations of interstellar gas and dust, stars in the process of forming, and planet atmospheres in deep space.

I. INTRODUCE

About George Carruthers: Dr. Carruthers was born in Cincinnati and grew up in Chicago. At an early age, he developed an interest in physics, astronomy, and space, which was encouraged by his father, a civil engineer. Dr. Carruthers earned his Ph.D. in aeronautical and astronautical engineering from the University of Ilinois in 1964. He was only 25 at the time. Ask students what qualities they think Carruthers had that enabled him to achieve so much at such a young age.

About Carruthers' Work: An expert in ultraviolet radiation, Dr. Carruthers invented the Far-Ultraviolet Camera/Spectograph. He has received numerous awards for his work.

In 1972 NASA presented him with the Exceptional Achievement Scientific Award. In 1987 he was named Black Engineer of the Year. Then, in 2003, he was inducted into the National Inventors Hall of Fame, which recognizes pioneers in the aviation and aerospace industries. Carruthers is currently a senior astrophysicist at the Naval Research Laboratory in Washington, D.C.

2. DISCUSS

Use the copy master to prompt discussion. Some of the students may not agree that it is important to study objects in deep space. They might believe that it is wasteful to spend time and money investigating space when there are so many problems to solve on Earth. Encourage debate among students who have differing viewpoints on the topic.

Answers to Explain It:

1. Answers will vary. Possible answer: Studying Earth's atmosphere can give us information that will help us solve problems, such as air pollution, on Earth. Studying objects in deep space can give us clues to how planets and stars form.

2. Answers will vary. Possible answer: Dr. Carruthers was a pioneer because he was the first person to invent a camera capable of capturing pictures of objects in deep space. His induction into the National Inventors Hall of Fame shows that other people also considered him to be a pioneer.

3. EXTEND

After students have completed the Explore It, ask them to share what they have learned. Help students locate websites that feature photographs taken with Dr. Carruthers' cameras.

Career Focus: Suppose you are one of NASA's leading scientists. You have been asked to assemble a team to go with you to visit a space station. Who would you take on this mission? Why?

Clarence Elder

AT A GLANCE

born 1935
Georgia

education
Morgan State College

occupation
Inventor

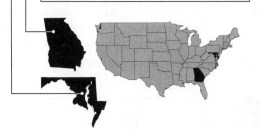

Energy resources are very important to us. They heat and cool our homes. They run our cars and machinery. But there is a limited supply of these resources. Our supplies of oil, coal, and natural gas will not last forever. This is why it is more and more important to conserve energy.

Clarence Elder knows a lot about conserving energy. In 1976, he invented the Occustat. This device helps reduce the amount of energy used in buildings. How does it work? Light beams are positioned at each door. The beams count people as they come and go. The lights tell the system to lower the heat when a room or building is empty or has just a few people in it. The Occustat can increase the heat when a lot of people are there. It can also control air-conditioning and lights.

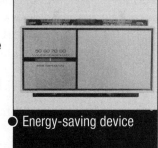

Energy-saving device

People all over the world benefit from Elder's invention. Many schools, hotels, and businesses use his system because it saves them money. And it greatly reduces the amount of energy used.

Elder makes the products he invents at his own company in Baltimore, Maryland.

? explain it

1. Why is it important to conserve energy?
2. What are some of the ways that you can conserve energy?

! explore it

Learn about the heating and cooling system in your school. Interview the head of your school's maintenance department. Ask: Does our school have an Occustat or similar system? What can students do to conserve energy?

Clarence Elder
TEACHING NOTES

Purpose: To introduce students to the inventor of the Occustat, an energy-saving system used around the world.

Science Background: Clarence Elder's Occustat was one of the first "smart-room" devices. Today, the Semantic Web promises to take smart rooms to a new level. The Semantic Web, which is still in development, is an extension of the World Wide Web. It is a markup language that will help move data more efficiently and help computers operate a wide range of electronic devices. Researchers have used a Semantic Web framework to develop a smart room in which a handheld computer controls a variety of devices. This is the first time a common computer language has been used to control numerous electronic devices.

The uses for such a technology are far-reaching. Government agencies, for example, could use Semantic Web to ensure that energy conservation policies are adhered to in all of their facilities. Consumers will also have many applications. A ringing phone, for example, might send a signal to a DVD player to stop a movie while the phone is answered.

1. INTRODUCE

About Clarence Elder: Elder studied electronics at Morgan State College (now Morgan State University) in Baltimore, Maryland.

About Elder's Work: Elder received a patent for his Occustat control system in 1976. Besides the Occustat, the inventor's research and development company, Elder Systems Incorporated in Baltimore, has developed more than 12 other patented devices. These include a timing device, a noncapsizable container, and a sweepstakes programmer. Elder has received several awards for innovations in electronics.

In 1965 and 1969 the New York International Patent Exposition awarded him a plaque for "Outstanding Achievement in the Field of Electronics."

Discuss with students the timeliness of Elder's invention: He invented the Occustat in the midst of an energy crisis that followed the Arab Oil Embargo in 1973. The rise in oil prices that followed the embargo contributed to the economic woes that plagued the nation throughout the 1970s.

2. DISCUSS

Use the copy master to prompt discussion. Help students understand the concept of a non-renewable resource. Explain the consequences of depleting a resource such as petroleum. Considering these consequences, help students understand that it is vital for everyone to conserve energy.

With the class, start a list of energy conservation tips on the board. These should be tips that everyone can follow on a daily basis.

Answers to Explain It:
1. Answers will vary. Possible answer: If we do not conserve energy we will run out of nonrenewable resources such as petroleum, natural gas, and coal.

2. Answers will vary. Possible answer: You could take shorter showers and turn off lights in vacant rooms. In the summer you could close drapes or shades to block the sun, therefore helping the air-conditioner run more efficiently.

3. EXTEND

After students have completed the Explore It, ask them to share what they have learned. Discuss computers in relation to current or future energy conservation technologies.

Career Focus: You have been asked to make a device that will save energy in your community. How will your device work? What are its benefits?

Jonathan Farley

AT A GLANCE

born 1970
Rochester, New York

education
A.B., Harvard University
Ph.D., Oxford University

occupation
Mathematician

Jonathan Farley comes from a family of high achievers. As a math major at Harvard, he earned high honors. As a result, he was invited to study math at Oxford University in England. While in England, Farley won top awards for mathematics research. He also earned his Ph.D. in mathematics.

When he returned to the United States, Dr. Farley began teaching and researching math. He worked on two math problems that had puzzled mathematicians. In 1998 he solved one of these problems.

○ **FARLEY's** math theories have won awards.

The next year he solved the other one.

Besides his interests in math, Farley also likes to write. He started a company to help Hollywood writers include math in their movies.

Farley received the Harvard Foundation's Distinguished Scientist of the Year Award in 2004 for his research.

? explain it

1. Did Dr. Farley's family have anything to do with his success? Why?
2. Based on what you know about Dr. Farley, what do you think his future holds?

! explore it

Jonathan Farley attended Harvard and Oxford, two of the most respected schools in the world. Find out more about one of these schools. Who are some of these schools' most famous students? What did they accomplish?

Jonathan Farley
TEACHING NOTES

Purpose: To introduce students to a mathematician who solved two mathematical problems that had been unsolved for many years.

Mathematics Background:
Dr. Jonathan Farley spends much of his research time dabbling in lattice theory and the theory of ordered sets (combinatorics). Lattice theory, a branch of abstract algebra, is the study of lattices, or sets of objects. At its most basic level, combinatorics is the study of counting and groups of numbers. Mathematicians who work in combinatorics examine how numbers can be ordered and the characteristics of these finite sets. The ways these sets can be combined is also a feature of combinatorics.

I. INTRODUCE

About Jonathan Farley: Dr. Farley is the quintessential high achiever. He graduated summa cum laude from Harvard University in 1991. He was one of only four Americans to win the Fulbright Distinguished Scholar award to the United Kingdom in 2001–2002. He earned a Ph.D. in mathematics from Oxford University. While at Oxford, he won the Senior Mathematical Prize and Johnson Prize, the university's highest mathematics awards, for his research. After returning to the United States, he held a fellowship at the Mathematical Sciences Research Institute in Berkeley, California.

About Farley's Work: Dr. Farley is a professor in the Department of Applied Mathematics at MIT. He is currently a Science Fellow at Stanford University's Center for International Security. His research is concentrated in the areas of lattice theory and the theory of ordered sets. In 2004 the Harvard Foundation presented Dr. Farley the Distinguished Scientist Award of the Year in recognition of his contributions in mathematical research.

2. DISCUSS

Use the copy master to prompt discussion. Help students understand that families can have a great amount of influence on a person and his or her success in a given career. Nonetheless, some people become successful with little or no backing from family members. Discuss the possible reasons why this is true.

Answers to Explain It:
1. Answers will vary. Possible answer: Jonathan Farley's mother, father, and brothers were all high achievers. They achieved academic as well as professional success in a number of areas. Their success probably had a big influence on Farley.

2. Answers will vary. Possible answer: Dr. Farley has accomplished a lot so far. He has demonstrated that he enjoys learning and achieving. It is highly probable that he will continue to achieve great things in mathematics and perhaps other areas such as writing.

3. EXTEND

After students have completed the Explore It, ask them to share what they have learned. Students might like to write short biographical profiles of one or more of the people they discover during their research.

Career Focus: You are interested in a career in mathematics. In what ways do you think Dr. Jonathan Farley would serve as a good role model?

Evan B. Forde

AT A GLANCE

- **born** 1952
  Miami, Florida
- **education**
  B.S., M.S., Columbia University
- **occupation**
  Oceanographer

Are you interested in what's going on under the sea? Evan B. Forde is. As a child, he loved watching "The Undersea World of Jacques Cousteau" on TV.

Science became his special interest. When he was in third grade, he owned a microscope, a telescope, and a chemistry set. His father, a science teacher, encouraged him.

Forde studied marine geology at Columbia University. While still a student, he began working at the National Oceanic and Atmospheric Administration as a researcher, and

FORDE conducts tests in a small submarine.

he's still there. He was the first African-American oceanographer to explore the ocean in a small submarine.

Forde is an expert in submarine canyons, large canyons that form beneath the ocean. He also studies underwater landslides that can cause tsunamis, giant waves of water that can destroy coastal areas.

? explain it

1. Why do you think Forde developed an interest in science at such a young age?

2. Why is it important to study features beneath the ocean?

! explore it

Find out more about submarine canyons. Draw a diagram that shows a submarine canyon in relation to other features beneath the sea.

Evan B. Forde
TEACHING NOTES

Purpose: To introduce students to an oceanographer/marine geologist who has created a successful career at the National Oceanic and Atmospheric Administration (NOAA).

Science Background: The ocean floor is divided by scientists into four main areas. The continental shelf extends from the shore into the ocean. The continental slope extends from the edge of the continental shelf to the part of the ocean floor that dips down steeply. At the base of the continental slope is the continental rise where sediment piles up from rivers flowing into the ocean. The large, flat areas away from the shore are known as abyssal plains.

Submarine canyons are deep canyons that occur on the continental shelf and slope. They often run perpendicular to seacoasts. There are several theories to explain the formation of these canyons. One theory suggests that these canyons were formed at a time when the oceans were much lower than they are now. The lower sea level permitted rivers to flow to the edge of the shelf. The erosional forces of the rivers then gouged out the canyons. Another theory suggests that the submarine canyons were carved by turbidity currents. Turbidity currents are currents that are produced by earthquakes. These currents move very quickly and carry sediment and debris along the ocean floor. Turbidity currents may be responsible for creating huge undersea landslides, sometimes causing large portions of the seafloor to move suddenly. Scientists theorize that these quick movements can trigger tsunamis.

1. INTRODUCE

About Evan B. Forde: Forde received his bachelor's degree in geology and his master's degree in marine geology from Columbia University. He joined NOAA in 1973 while still a student at Columbia. In 1979 he became the first African-American oceanographer to dive in a research submarine.

About Forde's Work: Forde is an expert on the formation and sedimentary processes of submarine canyons on the East Coast. He currently works at NOAA's Atlantic Oceanographic and Meteorological Laboratory in Miami, Florida. He is the former head of NOAA's Oceanic and Atmospheric Research Educational Outreach Committee. Over the years, he has spoken to thousands of students about oceanography and science in general. He has developed science experiments and wrote articles for *Ebony Jr.* magazine as well.

2. DISCUSS

Use the copy master to prompt discussion. Students should realize that Forde had a strong family support system. His family exposed him to science at an early age and then encouraged and fueled his interests through the years.

Help students understand that knowledge about submarine canyons and other underwater features can help scientists predict catastrophic events such as tsunamis.

Answers to Explain It:
1. Answers will vary. Possible answer: Forde's father was a science teacher. He probably was responsible for encouraging Forde's interest in science.

2. Answers will vary. Possible answer: Knowledge about the features below the ocean can help us understand why and when tsunamis occur. This can save lives.

3. EXTEND

After students have completed the Explore It, ask them to share what they have learned. Share with students the information from the Science Background section on this page. Encourage students to collect facts about other interesting underwater features.

Career Focus: You have always wanted to become an oceanographer. What are some of the things you would most enjoy doing? What is one aspect of your job that you might not like?

George F. Grant

AT A GLANCE

lived 1847–1910
Oswego, New York

education
D.D.S., Harvard University

occupation
Dentist, Inventor

Inventors are always thinking. They see a problem and they set out to solve it. George Grant thought about inventions while he worked as a dentist and even while he played golf.

Dr. Grant graduated from Harvard Dental School in 1870 and was hired to work there. He specialized in treating people who had a cleft palate, a birth defect on the roof of the mouth. Dr. Grant created a special device inserted into the mouth, which helped many people improve their quality of life.

GRANT's golf tee led to the invention of the modern tee.

When Dr. Grant was not working, he was usually on the golf course. Back then golfers did not put their ball on a tee before hitting it. Instead, they made a small mound of dirt or sand with their hands. Grant invented a wooden tee with a rubber top on which a golfer could place his or her ball. His tee helped lead to the invention of the modern tee.

? explain it

1. Which of Dr. Grant's two inventions was the most important? Why?
2. What role does curiosity play in the life of any inventor?

! explore it

Dr. Grant's invention greatly helped patients with cleft palates. Find out what doctors currently do to help people who are born with cleft palates.

George F. Grant
TEACHING NOTES

Purpose: To introduce students to a dentist who invented a prosthetic device for people with cleft palates. He also invented one of the earliest golf tees.

Science Background: Dr. George F. Grant was a mechanical dentist who specialized in helping people with cleft palates. Cleft palates develop in children during the early stages of pregnancy. The palate is formed from tissues on either side of the tongue. These tissues normally grow toward each other and join in the middle. For reasons yet unknown, sometimes these tissues fail to join, leaving a gap in the roof of the mouth. A cleft palate can cause problems with speaking and eating. In addition, it can affect the growth of the jaw and the development of teeth. Grant's invention, the *oblate palate*, was a prosthetic device that helped patients talk and eat better. Today, cleft palates are treated surgically. Typically, the surgery, which closes the gap in the palate, is performed when the child is 6 months old. Often, additional surgeries are necessary as the child grows. Sometimes a bone graft is done to improve the function of the palate.

1. INTRODUCE

About George F. Grant: Dr. Grant, the son of former slaves, was one of the first African Americans to graduate from the Harvard Dental School. After graduating, the school hired him as a faculty member. He remained on the staff for 19 years.

About Grant's Work: In dentistry, Grant is most known for the oblate palate. He gained international recognition for his invention. Grant's other invention—an early golf tee—helped improve his golf game and that of his immediate friends. His tee never became popular, however, because he never marketed it.

2. DISCUSS

Use the copy master to prompt discussion. Many students will say that the oblate palate was Dr. Grant's most important invention, although some may be inclined to say that the golf tee was more important. Encourage debate over whether a cosmetic medical procedure can be compared to a form of recreation.

Help students understand that curiosity is the driving force behind all innovative products and scientific advancements. Discuss the question that all inventors have ingrained in their minds: "What if…?"

Answers to Explain It:
1. Answers will vary. Possible answer: His most important invention was the device used to help people with cleft palates. This device helped many people live more normal lives.

2. Answers will vary. Possible answer: Curiosity is a trait all inventors share. Curiosity compels people to find out about things—how things work or why things are what they are. Through knowledge they can then find solutions to a problem or create a product that helps fill a need.

3. EXTEND

After students have completed the Explore It, ask them to share what they have learned. Share with students the information in the Science Background on this page. Help them locate websites that will provide them with information about current treatments for cleft palates.

Career Focus: If you had the choice of inventing something that would help people play a sport better or inventing something that would help people with a medical problem, what would you choose? Why?

Frederick **Jones**

AT A GLANCE

- **lived** 1893–1961
  Covington, Kentucky
- **education**
  Informal
- **occupation**
  Inventor

Lots of people have great ideas. Frederick Jones turned his ideas into useful inventions. During his long career, he invented more than 60 products.

One of Jones' best inventions began with a simple conversation he had with a truck driver in the 1930s. The truck driver delivered meat to stores. The man told Jones that he had lost an entire shipment because it had spoiled. The ice that cooled the meat had melted. Jones wondered if he could invent a new way to keep foods from spoiling during shipping.

JONES working on the details of an invention

Jones was naturally curious. He was also very good with mechanical things. After his talk with the truck driver, he came up with an idea that worked. He invented a small refrigeration unit called the Thermo King. The Thermo King, which was placed on trucks, could keep truck trailers cool. It could also be used in ships and railway cars to ship fresh foods.

? explain it

1. What two factors helped Jones invent the Thermo King?
2. How does Jones' refrigeration unit affect you?

! explore it

Frederick Jones invented many other products. Use the Internet to find out about some of these inventions. Which invention do you think was the most useful?

Frederick Jones
TEACHING NOTES

Purpose: To introduce students to an inventor of a portable refrigeration unit and many other useful products.

Science Background: Many foods will spoil if they aren't refrigerated. Each type of food spoils in its own way and is affected by specific types of *bacteria* and *fungi*. Chicken and other types of meat spoil in two ways: in the presence of oxygen and with no oxygen. If you take a piece of raw chicken out of the refrigerator, you may find that it is coated with slime. This slime is the product of bacterial growth in the presence of oxygen. If your piece of chicken has bacteria throughout it, it may also have *anaerobic* bacterial growth (bacteria that grow only where there is no oxygen). This type of spoilage causes unpleasant odors due to sulfides that are released as a by-product of the bacteria.

I. INTRODUCE

About Frederick Jones: Jones was born in 1893. He grew up an orphan and went to school only through the eighth grade. As a teenager, he supported himself by working in a garage. Although he originally was hired to sweep and clean the garage, his mechanical aptitude helped him become foreman of the car garage at the age of 15. Eventually he began designing and building race cars. When World War I began, he enlisted in the military. Jones began inventing things shortly after the war.

About Jones' Work: Jones patented more than 60 devices during his career. More than 40 of these devices involved refrigeration in some way. His automatic refrigeration system for long-haul trucks was patented in 1940. Jones was widely recognized for his work and was posthumously awarded the National Medal of Technology, a top honor for inventors. He was the first African American to receive the medal.

2. DISCUSS

Use the copy master to prompt discussion. Lead students to conclude that because food can now be shipped safely, we have a much wider variety of foods from which to choose. Ask students how many of them have eaten a piece of fruit today. Make a list of the fruits on the board. Discuss where each fruit comes from, and mark the ones that do not grow in your area. This exercise should give students a better understanding of the importance of Jones' invention.

Answers to Explain It:
1. Answers will vary. Possible answer: Jones was naturally curious. This curiosity helped him think of various ways to solve the problem presented. In addition, he was good with mechanical things.

2. Answers will vary. Possible answer: I can eat foods that are grown in faraway places. For example, I can eat fruit that was grown in South America. Refrigeration helps keep the fruit fresh for a long period. In addition, I can eat a wider variety of foods. Before Jones' invention, some foods could never be shipped because they spoil too quickly.

3. EXTEND

After students have completed the Explore It, ask them to share what they have learned. Students will discover that Jones invented a portable X-ray machine, an automatic ticket-dispensing machine, and many inventions related to refrigeration.

Career Focus: Think about a problem that exists in your school or community. What type of product or device would solve such a problem? How would you go about making your invention?

Ernest Everett Just

Dr. Ernest Just was 17 when he left South Carolina with only five dollars and headed for the Northeast.

In New Hampshire, he went to high school at Kimball Union Academy. He was a good student, and graduated first in his class in 1903.

Dr. Just's main interest was marine animals, or animals that live in the oceans. He went on to graduate school and earned his Ph.D.

Dr. Just spent most of his life studying and teaching. He spent 20 summers doing research on how marine animals

JUST studied how marine animals reproduce.

reproduce at Woods Hole Marine Biological Laboratory in Massachusetts. He also did research at institutions in Germany, Italy, and France. Scientists worldwide knew about his work from his books and articles.

For many years, Dr. Just taught biology and mentored students at Howard University in Washington, D.C.

lived 1883–1941
Charleston, South Carolina

education
Dartmouth College
Ph.D., University of Chicago

occupation
Biologist, Professor

? explain it

1. Why is it important to study how marine animals reproduce?
2. How might an oil spill affect the reproduction of marine animals?
3. In one sentence, how would you describe Dr. Just?

! exlore it

Dr. Just's work was done mostly on the eggs of marine invertebrates. Find out what an invertebrate is. Why is it easier to research the reproduction of animals that lay eggs rather than animals that give live birth?

Ernest Everett Just
TEACHING NOTES

Purpose: To introduce students to a scientist who was one of the first African Americans to receive a Ph.D. in science and who was a noted researcher in the field of marine invertebrate embryology.

Science Background: Marine invertebrate embryologists often work with sea urchin eggs (although Just worked with *Nereis*, a marine worm). Studying marine invertebrate eggs offers several advantages over studying other types of eggs: They are produced in large numbers and are fertilized externally, making them easy to study in the laboratory; both the egg and the embryo are transparent, so processes are easily observed in living tissue by using a microscope; and the developmental process is rapid (about 48 hours).

1. INTRODUCE

About Ernest Everett Just: Just was a pioneer in the sciences. He was one of the first 13 African Americans to receive a Ph.D. in the sciences. His desire was to be accepted as a scientist, rather than as an African-American scientist. For this reason, he temporarily left the United States in 1929 to work in Europe.

Adolf Hitler's rise to power in Germany in 1933 forced Dr. Just to leave Germany. In 1940 Just was in France when he was captured by German soldiers. He spent a short time in a prisoner-of-war camp until he negotiated his release. He then fled to Spain, Portugal, and back to the United States.

About Just's Work: Just concentrated his research on the development of the eggs of marine invertebrates. His first paper won him worldwide acclaim.

Just realized that the medium in which marine invertebrates live is vital to their reproductive process. Although his work took place before the advent of large shipments of oil across ocean waters, his understanding of the delicacy of the marine environment has helped other scientists understand and cope with changes in ocean waters due to oil spills.

In 1997, the United States Postal System released a postage stamp commemorating the life and work of Ernest Just.

2. DISCUSS

Use the copy master to prompt discussion. Lead students to conclude that certain habits of mind (qualities) such as those Just possessed are important to everyone. Point out that Just was 20 years old before he graduated from high school. Despite his late start, he achieved the highest honors in his field, including the first Spingarn Medal in 1915.

Answers to Explain It:
1. Answers will vary. Possible answer: The more we know about animals, the more we can do to help them. If a species is endangered, knowledge about its reproductive habits could prevent the species from becoming extinct.

2. Answers will vary. Possible answer: The oil spill could make the marine animals sterile. Or they might produce offspring that have abnormalities.

3. Answers will vary. Possible answer: Dr. Just was a dedicated scientist who studied long and hard to find out more about marine animals.

3. EXTEND

After students have completed the Explore It, ask them to share what they have learned. Discuss the fact that studying eggs is much easier than working with animals that give live birth because you can work with them in laboratories.

Career Focus: Scientists such as oceanographers study and research information about animal life found in the ocean. What are some animals that make their homes in the ocean? Would you be interested in studying marine animals? Why or why not?

Ronald E. McNair

AT A GLANCE

lived 1950–1986
Lake City, South Carolina

education
B.S., North Carolina A&T State University
Ph.D., MIT

occupation
Laser physicist, Astronaut

Ronald McNair grew up in a poor South Carolina town. But he never let this get in the way of his dream of being a scientist.

McNair graduated first in his high school class. When he was 26, he earned a Ph.D. in laser physics. Dr. McNair worked for a research lab in California and became an expert in lasers. He researched ways to use lasers on satellites.

Dr. McNair joined the space shuttle program in 1978. His job was to be a mission specialist, someone who performs experiments in space.

DR. MCNAIR at work

In 1984 Dr. McNair made his first flight in space. He was the second African American to do so. The mission was a success. McNair helped put two satellites into space.

Everything seemed to be going right for Dr. McNair. Then tragedy struck. In 1986 the space shuttle Challenger exploded, and McNair and six other people died.

Dr. McNair had wanted to play the saxophone on board the Challenger. It would have been the first piece of music played in space. The piece had been called "Rendezvous VI." But it was renamed "Ron's Piece" after his death.

? explain it

1. What traits helped Ronald McNair become a successful scientist?

2. Besides his interests in science, Dr. McNair played the saxophone and was an expert in karate. What do these facts tell you about him?

! explore it

Ronald McNair was a specialist in lasers. Find out about lasers. What do people use them for?

Ronald E. McNair
TEACHING NOTES

Purpose: To introduce students to a laser physicist and astronaut who lost his life aboard the space shuttle Challenger.

Science Background: Although the word laser is commonly used as a noun today, it is actually an acronym that stands for "light amplification by stimulated emission of radiation." A laser is an amplified, intense beam of coherent light. This light is photons, individual particles of energy released when electrons move within an atom. Coherent lights are the photons of a single frequency, or a combination of a few frequencies moving in one precise direction. This is opposed to incoherent light, which is a mixture of photons going in many different directions.

Lasers were first developed in 1960. They are now widely used and have a variety of applications. In medicine, they are used in cosmetic and eye surgeries. In industry, lasers are used to cut metals and other materials. Class IV lasers, such as the CO_2 laser, can cut through steel an inch thick. Lasers appear in numerous consumer products and devices. These include bar code scanners, printers, and compact disc players.

development of lasers and researched laser modulation for satellite space communications. As a mission specialist for NASA, McNair participated in a successful space shuttle mission in early 1984. Among other things, the mission successfully deployed two Hughes communications satellites.

2. DISCUSS

Use the copy master to prompt discussion. Students should conclude that McNair was a very determined individual. He apparently enjoyed setting and achieving lofty goals. His involvement and achievement in areas other than science—music and karate, for instance—show that he was a well-rounded person. He was a high achiever who realized the importance of balancing many activities, pursuits, and interests.

Answers to Explain It:

1. Answers will vary. Possible answer: McNair was a determined, goal-oriented person. He treasured the learning experience, which was sparked by his curiosity about how things work.

2. Answers will vary. Possible answer: McNair was a well-rounded individual who had many interests. He may have been a perfectionist because of the high degree of success that he achieved in a variety of activities. He enjoyed challenges.

1. INTRODUCE

About Ronald McNair: Ronald McNair was the quintessential high achiever. He was the valedictorian at Carver High School. He graduated magna cum laude from North Carolina A&T. He was named a Presidential Scholar; he received his Ph.D. in laser physics at the age of 26. He was a black belt karate instructor, and an accomplished jazz saxophonist.

About McNair's Work: After graduating from MIT in 1976, McNair went to work as a staff physicist at Hughes Research Laboratories in Malibu, California. While at Hughes, he worked on the

3. EXTEND

After students have completed the Explore It, ask them to share what they have learned. Share with students the information in the Science Background section on this page. Some students might enjoy making time lines showing the major developments in the history of the laser.

Career Focus: Would you ever consider being a mission specialist for NASA? Why? What would be some of the positive aspects? What would be some of the negatives?

Garrett A. Morgan

AT A GLANCE

lived 1877–1963
Paris, Kentucky

education
Elementary school

occupation
Inventor

Garrett A. Morgan was a son of former slaves. As a teenager, he left Kentucky and moved to Cincinnati to find a job and further his education.

Morgan hired someone to teach him English grammar so he could get a good job.

He became a successful business owner and inventor. Morgan worked in a Cleveland sewing shop repairing sewing machines. Then he opened his own sewing equipment and tailoring shop. His employees sewed clothes on equipment Morgan had made.

Traffic jams were common before Morgan's traffic light.

One day he saw a bad accident that inspired him to invent a traffic signal to help people cross the street safely.

Morgan became a national hero by using a gas mask he invented. In 1916 he used his gas mask to rescue men who were trapped in a tunnel under Lake Erie. The U.S. Army used his gas mask design in World War I.

? explain it

1. How did Garrett Morgan's inventions help people?
2. How did working in a sewing shop help Morgan later in his life?
3. Did Morgan value education? Explain your answer.

! explore it

Seach the Internet to find more information about how Garrett Morgan invented the gas mask and became a national hero shortly after.

Garrett A. Morgan
TEACHING NOTES

Purpose: To introduce students to an inventor who developed several important, life-saving devices.

Science Background: During World War I, Germany used mustard gas, chlorine, and other chemical cocktails in an attempt to gain an advantage in the war. The Allies reciprocated by using poison gases of their own against the Germans.

At first, only primitive protection devices were available to the soldiers. Prior to 1915, soldiers were given simple wads of cotton that were soaked in a solution containing baking soda. These pads were held over the face. Others used cloths drenched in human urine to provide protection.

A much better defense came toward the end of the war when Allied forces began using a version of the gas mask invented by Garrett Morgan. The masks, which used charcoal or antidote chemicals to filter the poison gases, were highly effective. The use of poison gases in warfare was outlawed in 1925.

I. INTRODUCE

About Garrett A. Morgan: The son of former slaves, Morgan worked on the family farm during his youth. As a teenager, he left Kentucky for Cincinnati, Ohio, looking for an education and a career. Despite having only an elementary school education, Morgan serves as a good role model for students. Emphasize to students the fact that he realized the importance of having a knowledge of English grammar. He hired a tutor to teach him grammar while he lived in Cincinnati.

About Morgan's Work: One of Morgan's most important inventions was the Morgan traffic signal. Although other traffic signals were invented and marketed before his, he was the first to patent such a device. His traffic signal was a T-shaped pole that had three positions: stop, go,

and another stop that halted traffic in all directions and allowed pedestrians to safely cross the street. It could be found throughout North America and Britain until the development of the familiar red, yellow, and green signals used today. Morgan sold the rights to his invention to General Electric for $40,000.

2. DISCUSS

Use the copy master to prompt discussion. Help students understand that two of Morgan's inventions helped save an unestimatable number of lives. His traffic signal most likely prevented many accidents. A version of his gas mask also saved the lives of many soldiers who used them to protect themselves from deadly gases used by Germany during World War I.

Answers to Explain It:
1. Answers will vary. Possible answer: Morgan's gas mask helped save the lives of soldiers during World War I. His traffic signal saved the lives of pedestrians as well as automobile passengers.

2. Answers will vary. Possible answer: He probably learned a lot about sewing machines and how they work while working in the sewing shop. He used this knowledge to invent a zigzag attachment for sewing machines.

3. Answers will vary. Possible answer: He valued an education because he hired a tutor to teach him English grammar while he was in Cincinnati.

3. EXTEND

After students have completed the Explore It, ask them to share what they have learned. Students should relate the story of how Morgan used his gas mask to rescue several men who were trapped in an underground tunnel beneath Lake Erie in 1916. Morgan gained nationwide attention after the daring rescue. Fire departments eagerly sought the device and the U.S. Army refined the design for use during World War I.

Career Focus: Think about a safety product that you might invent. Who would be interested in purchasing your product? How could learning about Garrett Morgan inspire you to complete your project?

Katherine Adebola Okikiolu

AT A GLANCE

born 1965
England

education
B.A., Cambridge University
Ph.D., UCLA

occupation
Mathematician

Katherine Okikiolu's family has mathematics in their blood. Her father is a famous mathematician. Her mother is a high school math teacher.

Dr. Okikiolu also loves math. When she was young, she thought of math as being "mysterious" and "extraordinary."

Dr. Okikiolu earned her Ph.D. from UCLA. She impressed her teachers by solving some very difficult math problems. In 1997 she won an important award from the U.S. government. The award is given yearly to promising young math researchers. She also won a Sloan Research Fellowship.

OKIKIOLU studies sound pitches from drums.

Today Dr. Okikiolu teaches math in California. She is also studying the sounds that drums make and then putting these sounds into math formulas. And she is making a series of videos. Dr. Okikiolu hopes these videos will help inner-city children learn numbers and measurements in a fun way.

explain it

1. How did Dr. Okikiolu become interested in mathematics?
2. What evidence suggests that Dr. Okikiolu wants to share her interests in math?

explore it

Mathematicians do much more than create formulas and do computations. Use the Internet and talk to a math teacher to find out about the jobs available in the field of mathematics.

Katherine Adebola Okikiolu
TEACHING NOTES

Purpose: To introduce students to a promising mathematics researcher who works at MIT and UCSD.

Mathematics Background:

Okikiolu's research includes the study of elliptical determinants to geometry. This involves studying the properties of various dimensions in space. Dr. Okikiolu is currently researching two-dimensional drums. She is using mathematical formulas to determine if different drums can produce the same sounds. Likewise she is looking for patterns between frequencies and pitches. Okikiolu hopes her research will shed new light on three-dimensional drums and even drums of a higher dimension that cannot be constructed in our three-dimensional world. Such research may eventually be applied to problems in quantum physics. Okikiolu's work illustrates how mathematics can be applied directly to other disciplines. Math is more than adding numbers and solving equations. Science and math are used together in the development of new technologies. Mathematics also helps to improve problem-solving skills.

1. INTRODUCE

About Katherine Adebola Okikiolu: Okikiolu was born and raised in England. She earned her B.A. in mathematics from Newnham College, an all-women's college at Cambridge University. She moved to the United States in 1987, enrolled at UCLA, and earned her Ph.D. in 1991.

About Okikiolu's Work: After graduating from UCLA, Okikiolu served as an instructor and assistant professor at Princeton. In 1997 Okikiolu received one of the 60 Presidential Early Career Awards for Scientists and Engineers. The annual awards were established by President Clinton in 1996. She currently holds positions at MIT and UCSD. Her research interests include classical analysis, differential geometry, partial differential equations, and operator theory.

2. DISCUSS

Use the copy master to prompt discussion. Help students understand that Okikiolu's interests in math were cultivated and encouraged by her parents, who are both mathematicians. Discuss with students other ways that people get interested in particular subject areas or careers.

Answers to Explain It:

1. Answers will vary. Possible answer: Her parents helped her develop an interest in mathematics. Both of them had careers in math. In addition, she developed interests on her own by reading and studying math books in her spare time.

2. Answers will vary. Possible answer: She has developed a series of videos to help inner-city children learn math.

3. EXTEND

After students have completed the Explore It, ask them to share what they have learned. Invite a mathematician to class. If you are unable to arrange for a speaker, you might like to point students to various resources that describe the various jobs that are done by people with mathematics backgrounds. The following website provides a lot of useful information:

http://www.ams.org/careers/

Career Focus: You have been asked to calculate the distance between Earth and several other planets. How might this information be useful?

Waverly **Person**

Waverly Person knows a lot about earthquakes. When a large earthquake occurs, he is usually contacted by the media to get information.

This happened in the Indian Ocean in December 2004. The earthquake created a terrible tsunami that killed thousands of people. A tsunami is a series of large waves. These waves travel as quickly as an airplane and are as high as 30 feet.

As director of the National Earthquake Information Center, Person told reporters that "most of the people could have been saved if they had had a tsunami warning system in place." Such a system warns people when a tsunami is likely to occur. It gives them time to move away from the coasts. The United States has warning centers in the Pacific Ocean and in Alaska and Hawaii. Such systems need to be established in Asia, along with educating people on what to do.

A young Person at work.

AT A GLANCE

born 1927
Blackenridge, Virginia

education
B.S., St. Paul's College
American University
George Washington University

occupation
Geophysicist, Seismologist

? explain it

1. Why do the media call Waverly Person whenever an earthquake occurs?
2. What does Person think governments can do to help save lives?

! explore it

Waverly Person is the head of the U.S. Geological Survey's National Earthquake Information Center. Find out about this center. What is its mission?

Waverly Person
TEACHING NOTES

Purpose: To introduce students to a geophysicist and seismologist who was the director of the U.S. Geological Survey's National Earthquake Information Center.

Science Background: The United States Geological Survey (USGS) estimates that millions of earthquakes occur around the world each year. Most of these go undetected, however, because they have small magnitudes or occur in remote areas. The USGS maintains that earthquakes pose a "significant risk to 75 million Americans in 39 states."

I. INTRODUCE

About Waverly Person: Person attended St. Paul's College in Lawrenceville, Virginia. After a three-year assignment with the U.S. Army, he moved to Washington, D.C., where he performed a variety of jobs until obtaining a position with the National Earthquake Information Center in 1962. While employed there he attended American University and George Washington University, becoming a geophysicist in 1973.

About Person's Work: Person was the director of the National Earthquake Information Center. He has earned several awards. Among these is the Outstanding Government Communicator Award and the Meritorious Service Award from the U.S. Department of the Interior.

2. DISCUSS

Use the copy master to prompt discussion. Students should understand that tsunami warning systems, such as those in the United States, are not in place in Asia because large tsunamis are relatively rare in the Indian Ocean. Discuss how the lessons learned during the 2004 tsunami

in the Indian Ocean might encourage governments to implement warning systems. Ask students how people such as Waverly Person might be valuable resources to these countries.

Answers to Explain It:

1. Answers will vary. Possible answer: They call Person because he is an earthquake expert. He is the director of the National Earthquake Information Center.

2. Answers will vary. Possible answer: Person believes that lives could be saved if Asian nations installed tsunami warning systems like those in Hawaii and Alaska. In addition, he thinks that governments should educate people—especially those living along coasts—by telling them what to do in the event of an earthquake.

3. EXTEND

After students have completed the Explore It, ask them to share what they have learned. Direct students to the National Earthquake Information Center Website:

http://neic.usgs.gov/

Some students might enjoy periodically tracking earthquake activity around the world. The Web site has maps that show earthquake activity over the last seven days. Map areas showing activity can be clicked to get specific information.

Career Focus: Imagine you are a seismologist living in a nation bordering the Indian Ocean. What are some of the things you could do to educate the people living along the coast so that they are better prepared in the event of an earthquake?

Arlie **Petters**

AT A GLANCE

born 1964
Dangriga, Belize

education
B.A., M.A., Hunter College
Ph.D., Massachusetts Institute
of Technology

occupation
Mathematician

They say that beauty is in the eye of the beholder. To a young Arlie Petters, beauty was art. He loved to draw pictures of "distant horizons" and "mystical settings." He also found beauty in space and stars.

In high school, Petters also discovered that there was beauty in mathematics. He remembers discovering that math "has a beauty of its own." He said, "I discovered that the same joy I felt when I did art is present in…mathematics."

○ **PETTERS** uses math to understand starlight.

Dr. Petters has combined his interest in math and space into a fascinating career. He writes mathematical formulas that trace the pathways of starlight.

Besides his research, Dr. Petters enjoys teaching. He believes students have a natural fascination with the stars and what is beyond them. He believes he can spark their interest in math by combining it with the study of the stars.

? explain it

1. What are some of the things Dr. Petters considers beautiful?
2. What are some of the benefits of having a career that explores one or more of your interests?

! explore it

Part of Dr. Petters' research involves the study of quasars. Research the Internet to find out what quasars are and what they do.

Arlie **Petters**
TEACHING NOTES

Purpose: To introduce students to a mathematician who is an expert in gravitational lensing, a method of tracing intricate pathways of starlight.

Mathematics Background: Arlie Petters is credited with developing the first general mathematical theory of gravitational lensing. Gravitational lensing is the method of tracing the pathways of starlight. Lensing, or the distortion of light, occurs when powerful gravitational forces bend light as a lens does, often splitting a single image of a star into multiple images. Astronomers have long been interested in gravitational lenses because they help them understand the structures of objects such as distant galaxies, black holes, and stars. Yet they did not have a general mathematical theory to explain the lenses' properties. Petters helped fill this void. Now astronomers have a tool that they can use to help them analyze the many intricacies of starlight and understand what gravitational forces affected it on its way to Earth.

I. INTRODUCE

About Arlie Petters: Dr. Petters is considered one of the founders of the field of mathematical astronomy. He emigrated to the United States from Belize in 1979 and became a U.S. citizen in 1990. He holds bachelor's and master's degrees from Hunter College and a Ph.D. in mathematics from MIT. In the 1990s he was an instructor in pure math at MIT and a visiting mathematician at Oxford University's Mathematical Institute. He is currently a full professor at Duke University.

About Petters' Work: Petters is mainly interested in the mathematical theory of gravitational lensing and areas related to it, such as singularity theory and theoretical astrophysics. Singularity theory involves examining points and sets. Theoretical astrophysicists study everything in the universe from string theory to dark matter and dark energy. Petters has published 30 papers and one book, *Singularity Theory and Gravitational Lensing*. A course he teaches, "The Mathematics of Light Deflection in the Universe," is based on the book. The course aims to "teach students how regular calculus can be extended and applied to real problems in gravitational lensing."

2. DISCUSS

Use the copy master to prompt discussion. Help students understand that "beauty" is a term that means something different to everyone. Ask students what things they consider beautiful.

Discuss some of the drawbacks of working in a field that one dislikes.

Answers to Explain It:
1. Answers will vary. Possible answer: He considers art, math, and the stars beautiful.

2. Answers will vary. Possible answer: You would look forward to going to work every day. Stress probably wouldn't be a factor and therefore you would probably be healthier as a result.

3. EXTEND

After students have completed the Explore It, ask them to share what they have learned. You might like to refer students to websites that describe quasars.

Career Focus: What are some of the reasons why it might be interesting to work in a field such as the one Dr. Petters works in?

Barbara Ross-Lee

born 1942
Detroit, Michigan

education
B.S., Wayne State University
M.D., Michigan State University
College of Osteopathic Medicine

occupation
Osteopathic Physician

Barbara Ross-Lee grew up in Detroit, Michigan. Her family was poor, but it was rich in support. Why?

The Ross family loved music. They encouraged her sister's love of singing. Diana Ross became famous as the lead singer of a group called the Supremes.

Barbara liked music, too, but she wanted to become a doctor. Her family supported her choice.

Becoming a doctor was not easy, however. Her adviser said women should not be doctors. Ross-Lee's hopes of becoming a doctor faded.

ROSS-LEE specializes in muscles and bones.

But in 1969, Michigan State University started a new school for doctors. Ross-Lee trained as a special doctor, treating diseases by adjusting bones and muscles.

Dr. Ross-Lee started a private practice in 1973. Then she became the head of a large medical school. Dr. Ross-Lee was the first African-American woman to head such a school.

? explain it

1. What event tells you that Dr. Ross-Lee was a very determined person?
2. Barbara and her sister Diana were both successful. Why do you think they were so successful?

! exlore it

Dr. Ross-Lee is a doctor of osteopathic medicine (DO). Find out what a DO does. What schools in your area train people to become DOs? What classes would be helpful for a person who wants to become a DO?

Barbara Ross-Lee

TEACHING NOTES

Purpose: To introduce students to a successful osteopathic physician who had to overcome many hurdles, including poverty and sexism.

Science Background: According to the American Osteopathic Association (AOA), "doctors of osteopathic medicine (DOs) practice a 'whole person' approach to health care." Instead of just treating specific symptoms, osteopathic physicians treat the entire body at once.

A technique called osteopathic manipulative treatment (OMT) has long been a key part of osteopathic medicine. In this technique, physicians use their hands to diagnose. DOs feel for difference between muscles and tendons, suggesting restricted motion or chronic pain.

Osteopathic doctors, like MDs, are considered physicians. They are licensed by state and specialty boards. DOs can perform surgeries and write prescriptions. Training for MDs and DOs is very similar. Both types of doctors must obtain a four-year degree with an emphasis on the sciences. Then they must complete a three or four year medical program. Finally, before obtaining a position in a hospital or private practice, they must complete a residency program.

I. INTRODUCE

About Barbara Ross-Lee: Dr. Ross-Lee overcame much adversity to attain her goal of becoming a physician, making her a good role model for students. Growing up in a housing project in Detroit, Michigan, she often faced racial discrimination. Then, while attending Wayne State University, she encountered blatant sexism when her adviser told her she could not choose human anatomy as her major because women should not become doctors.

About Ross-Lee's Work: Ross-Lee graduated from the Michigan State University College of

Osteopathic Medicine in 1973. After graduation she developed a private practice in Detroit. In 1993 she was appointed dean of the College of Osteopathic Medicine of Ohio University. She served in that position until 2001. Ross-Lee has received numerous awards. In addition, she served on the National Advisory Committee on Rural Health of the United States Department of Health and Human Services.

2. DISCUSS

Use the copy master to prompt discussion. Discuss the role that determination plays in almost every success story. What other traits are usually associated with people who are high achievers? Talk with students about the importance of having a family support system. Is it easier to accomplish something if you have someone encouraging you to succeed?

Answers to Explain It:
1. Answers will vary. Possible answer: She was determined to become a doctor even though her adviser told her that women should not be doctors.

2. Answers will vary. Possible answer: In order to be successful, it is important to have a family that supports you. Ross-Lee's family supported and encouraged both Diana and Barbara.

3. EXTEND

After students have completed the Explore It, ask them to share what they have learned. In addition, you might like to share some of the information contained in the Science Background notes on this page.

Career Focus: What are some of the positive things about being a doctor? What are some of the negative things about being a doctor? Draw a line down the middle of your paper to make these two lists. Would you like to be a doctor? Why or why not? Be prepared to discuss your ideas with the class.

Reva Kay Williams

Deep space holds many mysteries. Astronomers such as Dr. Reva Williams devote their lives to uncovering them.

One of the biggest puzzles deals with black holes. Scientists think these are large regions in space that form when a giant star or group of stars dies or collapses. Gravity sucks in everything near it, even light particles.

Astronomers were puzzled why some black holes give off light. In 1969 Roger Penrose came up with an explanation. He said that light particles are made in a part of the black

○ Black holes do produce streams of light.

hole where gravity is the strongest. This force gives off the light, making circles of light around the black hole.

Dr. Williams proved Penrose's theory in 2004. Computer models she made show that black holes do produce streams of light. In addition to proving the theory, she also helped explain much about the appearance of the light.

AT A GLANCE

born
Memphis, Tennessee

education
B.A., Northwestern University
M.A., Ph.D., Indiana University

occupation
Astronomer, Astrophysicist

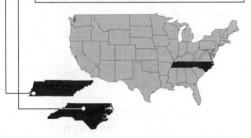

? explain it

1. Why do you think it is important to study black holes or other mysteries in space?
2. What did Dr. Williams prove?

! explore it

The knowledge we have about black holes is changing all the time. Use the Internet to find out about some of these new ideas.

Purpose: To introduce students to a successful astrophysicist who was the first person to prove the Penrose Mechanism theory.

Science Background: The Penrose Mechanism is a theory originally posed by Oxford mathematics and physics professor Roger Penrose in 1969. The theory says the rotational energy of a stellar black hole—which is caused by the collapse of stars—makes and emits high-energy particles, often spewing them great distances.

Dr. Williams' research supports and expands upon Penrose's theory. Using computer modeling, she showed that the high-energy particles—protons and electrons—are made within a portion of a black hole where the gravitational forces are so strong that they bend rays of light around the black hole.

Dr. Williams' research also offers an explanation for the one-sided appearance of the rays. She says the appearance is due to the gravitational dragging of space and time near the cores of a black hole. Previously, scientists believed the appearance was due to observers' position relative to the rays. According to Fernando de Felice, a physicist at the University of Padova in Italy, "The interest in Dr. Williams' work is that it has enriched the possibilities of having energy output in active cosmic sources."

I. INTRODUCE

About Reva K. Williams: Dr. Williams received her B.A. in astronomy and physics from Northwestern University. She earned her M.A. and Ph.D. (astrophysics) from Indiana University. She was the first African-American woman to gain a doctorate in astrophysics.

About Williams' Work: Dr. Williams currently teaches and conducts research at Bennett College and the University of Florida. Her research areas include relativistic astrophysics, general rel-

ativity, cosmology, and extragalactic astronomy. Dr. Williams says that she equally enjoys "doing research and teaching to share my knowledge."

2. DISCUSS

Use the copy master to prompt discussion. Discuss the fact that knowledge itself is an adequate reason to study something. Encourage students to present contrary arguments.

Talk with students about the ever-changing science of astronomy. Discuss how new information could and probably will alter the way scientists view the information discovered by Dr. Williams.

Answers to Explain It:
1. Answers will vary. Possible answer: Such studies might shed new light on the origins of the universe and help us better understand the universe.

2. Dr. Williams proved the theory posed by Roger Penrose. He believed that light particles begin in a section of a black hole and are then cast off.

3. EXTEND

After students have completed the Explore It, ask them to share what they have learned. Help students locate websites that feature grade-level information about black holes. Suggest that some students might like to locate and then print illustrations of black holes. These illustrations and collected facts could be used to construct a poster or a fact sheet.

Career Focus: What are some of the tools that are probably valuable to astronomers? What skills do all astronomers probably share?

Roger Arliner Young

AT A GLANCE

lived 1899–1964
Clifton Forge, Virginia

education
B.S., Howard University
M.S., University of Chicago
Ph.D., University of Pennsylvania

occupation
Zoologist

Do you enjoy learning about animals? Roger Arliner Young made it her career.

Dr. Young did not have an easy childhood. She had to care for her sick mother. Yet she went on to become an expert in zoology, the study of animals.

Young went to Howard University. There she took her first science course. Her teacher was Ernest Everett Just, a well-known biologist.

Young was a poor student at first, but Just encouraged her. Young's grades greatly improved. In 1923 she earned her bachelor's degree.

YOUNG studied marine animals.

In 1926 she earned a master's degree.

Young began working with Ernest Just. She described why cells gain or lose water. Just liked her work, and called her "a real genius in zoology."

After failing the required tests on her first try, Dr. Young reached her goal. In 1940, she became the first African-American woman with a Ph.D. in zoology.

❓ explain it

1. Why is it important to study animals?
2. Would you say that Young was a determined person? Why?
3. How did Ernest Just help Young?

❗ explore it

Find out about some of the things that zoologists do. Create a pamphlet that shows some of these jobs. Illustrate your pamphlet with photographs or drawings.

Roger Arliner Young
TEACHING NOTES

Purpose: To introduce students to a zoologist who became the first African-American woman to gain a Ph.D. in zoology.

Science Background: In the 1920s, Ernest Everett Just and Roger Arliner Young spent their summers at the Marine Biological Laboratory (MBL) at Woods Hole, Massachusetts. Many of their studies concerned the fertilization process in marine organisms. Today scientists continue to spend summers studying marine organisms there. Many of them concentrate on fertilization. Over the years, research on marine organisms has led to medical breakthroughs. For example, studies of the sea urchin conducted at the MBL helped lead to the development of test-tube fertilization.

The sea urchin has long been a favorite research subject for the following reasons: It is readily available in Woods Hole and is easy to maintain in the lab; female sea urchins produce as many as a half million eggs during the summer; the sea urchin's fertilization and embryonic development are rapid; and the sea urchin's development has been well documented by researchers, such as Young and Just, over nearly a century.

1. INTRODUCE

About Roger Arliner Young: Roger Arliner Young succeeded in a challenging field despite the fact that she had poor grades during her first year at Howard University. Ernest Everett Just was one of her science professors, and he encouraged her studies. Lead students to understand that people can achieve great things even if they start out slowly. Also talk about the importance of having someone encourage you when things are not going well.

About Young's Work: While at the University of Chicago, Young had her first research paper,

"On the Excretory Apparatus in Paramecium," published in a popular science magazine. Young worked closely with Ernest Everett Just at the Marine Biological Laboratory in Woods Hole, Massachusetts, in the late 1920s. There she helped Just with his research on the reproduction processes of marine organisms. She also conducted studies on water use in animal and plant cells.

2. DISCUSS

Use the copy master to prompt discussion. Lead students to understand that the study of animals helps us understand the special needs of different organisms. For example, young animals need special diets so that they can grow properly. Studying animals also helps us understand what causes diseases and how they may be treated or prevented. Studying animals enables us to explore parallels between animal and human diseases.

Answers to Explain It:
1. Answers will vary. Possible answer: Knowing about animals helps us protect them.

2. Answers will vary. Possible answer: She was a determined person because she earned a Ph.D. despite failing in her first attempt.

3. Just encouraged her. Because of his encouragement, she improved her grades.

3. EXTEND

After students have completed the Explore It, ask them to share what they have learned. Help students locate information about zoologists on the Internet. There are many career options. These include agriculturalist, animal trainer, entomologist, forester, fishery manager, science writer, and pathologist.

Career Focus: If possible, have a biologist or zoologist speak with the class about careers in zoology. What kinds of places employ zoologists?

Benjamin Banneker

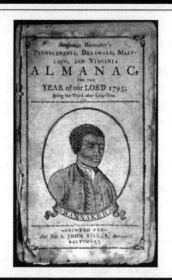

AT A GLANCE

lived 1731–1806
Baltimore, Maryland

education
Private community school

occupation
Astronomer

Benjamin Banneker was naturally curious. When he was 21, a man gave him a watch. He wanted to know how it worked. So he took it apart. He then carved wooden pieces similar to the watch pieces and fit them together. The result was a clock that ran for 40 years!

Banneker became interested in math and science at a small school near his family's farm, not far from Baltimore, Maryland.

In 1772 Banneker became friends with Joseph Ellicott. This friendship would change his life. Ellicott had an interest in astronomy. When Ellicott died, he left Banneker a telescope and books on astronomy. Banneker started to study the night sky and soon became a skilled astronomer. He predicted the time of a solar eclipse in 1789.

● **BANNEKER'S** Almanac was used across the country.

Banneker's predictions were more accurate than those of trained scientists. He began preparing a yearly book, *Banneker's Almanac*. It had information about eclipses, the weather, and the moon. It also had tables showing tides and the times of sunrise and sunset.

Joseph Ellicott's cousin, Andrew, employed Banneker as a surveyor. They worked as a team to plan what would become Washington, D.C.

? explain it

1. How did Banneker's curiosity help him?
2. How did Banneker's life change when he met Joseph Ellicott?
3. How was *Banneker's Almanac* useful to people?

! explore it

Benjamin Banneker was an important part of the team that designed and laid out Washington, D.C. Find out what it is about the layout of our capital that is special. Also determine what role Banneker played in the design.

Benjamin **Banneker**
TEACHING NOTES

Purpose: To introduce students to an historic figure whose focus on astronomy became the basis of an almanac used across the United States.

Science Background: Solar eclipses occur when the sun, the moon, and Earth align and the moon casts a shadow on Earth, blocking the sun from view. Eclipses can happen at two points during the yearly rotation cycle: the ascending node and the descending node. There are two eclipse seasons each year, during which times at least one eclipse occurs. This means that there are at least two eclipses each year. Sometimes Earth passes between the moon and the sun. A shadow is cast on the moon. This phenomenon is called a lunar eclipse.

Today, we can calculate the orbits of the three bodies involved in a solar eclipse by using a simple formula and a calculator. When Banneker made his predictions about solar eclipses, he had to take physical measurements of the positions of the sun, the moon, and Earth. He did calculations by hand, which took painstaking effort.

1. INTRODUCE

About Benjamin Banneker: Banneker, a freeman, wrote to Thomas Jefferson about racial injustice in America, a heated subject. Jefferson praised the copy of the 1791 edition of *Banneker's Almanac* that he received. Banneker was part of the survey team for the new capital city of Washington, D.C.

About Banneker's Work: Banneker made major contributions to his country and his community as well as to the worldwide scientific community. Banneker spent his retirement on his farm outside Baltimore, where he was often visited by other distinguished scientists. Upon his death in 1806, he was eulogized at the Académie des Sciences in Paris and in the Parliament in London.

2. DISCUSS

Use the copy master to prompt discussion.
Lead students to understand that Banneker lived at a time when few people had advanced education. His enthusiasm for learning led him far beyond the achievements of most people of the time, of any race, and he willingly used his intellectual talents for the betterment of the nation.

Answers to Explain It:
1. Answers will vary. Possible answer: Curiosity encouraged him to question things and then to find answers to those questions.

2. Answers will vary. Possible answer: He developed a lifelong interest in astronomy.

3. Answers will vary. Possible answer: People are better able to plan their activities.

3. EXTEND

After students have completed the Explore It, ask them to share what they have learned.
Students will learn, among other things, that:

• Banneker was appointed to a team of three to design Washington, D.C. Major Pierre Charles L'Enfant, a French officer who fought with Washington during the Revolutionary War, was in charge. The other member was Andrew Ellicott, the cousin of Banneker's friend, Joseph Ellicott.

• After two years, L'Enfant quit the project and took the plans with him to France.

• Banneker reproduced the plans from memory and he and Ellicott completed the project.

Career Focus: The architectural design of your school may or may not be interesting. If you were asked to design a school, would you model it after your school or make changes? Explain your answers.

Otis **Boykin**

AT A GLANCE

lived 1920–1982
Dallas, Texas

education
Fisk University
Illinois Institute of Technology

occupation
Inventor

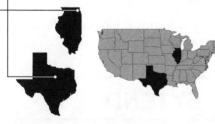

Have you ever heard of a pacemaker? These electronic devices help the heart beat regularly. Pacemakers have saved many lives.

Otis Boykin invented an electrical part for early pacemakers. This part controlled the pacemakers. He has also made many other useful devices. Some of them are still used today.

Boykin did not start out as an inventor. He first studied hard in school. He learned about technology. His parents were too poor to pay for a full college education. Despite not finishing school in Illinois, he got a job as a laboratory assistant testing controls for aircraft. But this work did not satisfy him, so he began inventing things.

He invented many parts for electronic devices. One of these

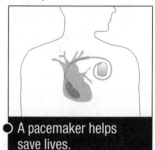

A pacemaker helps save lives.

parts was used in many radios, televisions, and computers. This part helped lower the cost of these items. Another device greatly helped the military. It was used to guide missiles to their targets.

Boykin shared some of his ideas with companies in France and the United States. As a result, many more useful products were created. He even worked with movie companies making sure the facts they presented were correct.

He died of heart failure in 1982.

? explain it

1. What was Boykin's most important invention? Why?
2. What qualities do you think are important to an inventor?
3. Have you ever had an idea for an invention? What was it?

! explore it

Many of Boykin's inventions were electronic devices called resistors. Find out what a resistor is. What are some of the uses for resistors?

Otis **Boykin**
TEACHING NOTES

Purpose: To introduce students to an inventor of several important devices, one of which has saved many lives.

Science Background: In some people, the *sinoatrial node*, which controls the contractions of the heart, does not function properly. Pacemakers use electrical impulses to regulate heartbeats. The first implantable pacemaker was used in 1960. Since then, the technology used to make pacemakers has improved tremendously. A typical modern pacemaker is about the size of a pager. It contains a battery and the electronic circuitry that runs the device. In addition, pacemakers have two thin electrical wires. These wires connect the device to the heart.

A surgeon implants the device below the skin in the shoulder. The wires are then threaded through a vein that runs to the heart. One of the wires is threaded to the right atrium of the heart and the other is threaded to the right ventricle of the heart. The pacemaker and wires are then programmed to analyze the heartbeat and electrically stimulate the heart when necessary.

Although not all pacemakers are programmed the same way, most are programmed so that the electrical wires can perform two functions. One function is sensory. The device detects if the sinoatrial node is supplying electrical impulses to the heart. Another function is to transmit electricity from the pacemaker's battery to the heart when needed.

In essence, the pacemaker supervises the heart. It ensures that the heart contracts at a rate adequate to pump blood throughout the body.

I. INTRODUCE

About Otis Boykin: One of Boykin's first jobs was delivering packages. He attended Fisk University in Nashville and the Illinois Institute of Technology. Eventually, he became a laboratory

assistant, testing airplane controls. This work helped him to become a successful inventor.

About Boykin's Work: Boykin is responsible for inventing 27 electrical devices, 11 of which received patents. Much of his work centered on the development of innovative resistors. He created a resistor that was used in a variety of electronic devices, including televisions, radios, and computers. He also invented a variable resistor used in guided missile systems.

2. DISCUSS

Use the copy master to prompt discussion.
Ask students to name several inventions they consider to be the most important inventions of the 20th century. List these inventions on the board and discuss their benefits. What makes a great invention? What traits do you think most inventors share? Otis Boykin is a strong, positive role model. Ask students what the consequences might have been had he decided to remain a laboratory assistant rather than turn his attention toward inventing things.

Answers to Explain It:
1. Answers will vary. Possible answer: Boykin's part for the pacemaker was his most important invention because it has helped save many lives.

2. Answers will vary. Possible answer: The person should have a good imagination and be a good problem solver.

3. Answers will vary. Encourage students to discuss their inventions openly.

3. EXTEND

After students have completed the Explore It, ask them to share what they have learned.
Help students understand the function of a resistor: It is an electrical conductor used to control voltage by its resistance. Have students write some uses of resistors on the board and discuss which are the most useful to them.

Career Focus: Think about some device that you might invent. How could learning about Otis Boykin help inspire you to complete your project?

Marjorie Lee **Browne**

AT A GLANCE

lived 1914–1979
Memphis, Tennessee

education
B.S., Howard University
M.S., Ph.D., University of Michigan

occupation
Mathematician, Professor

Marjorie Lee Browne grew up during the Great Depression. From 1929 to 1940, most Americans lost all of their money and investments in a period that came to be known as the Great Depression. Because of the nation's poor economy, many people were out of work.

Browne's family was lucky. Her father and stepmother had good jobs. They were able to provide her with an excellent education. While she was in high school, Browne won the Memphis city singles tennis championship.

Math was Browne's favorite subject, though. Her parents had a love for math and passed that interest on to Browne. She soon earned math degrees from two well-respected universities.

DR. BROWNE enjoyed teaching math.

During the 1940s, Browne taught math at several schools. In 1939 she started work on her Ph.D. Ten years later, after going to school part-time, she earned a Ph.D. in mathematics. She was one of the first African-American women to earn a doctorate in math.

Dr. Browne received a major award in mathematics education. After she retired, she helped many minority students with their math educations. Dr. Browne also gave students money out of her own pocket.

? explain it

1. What event in Dr. Browne's life showed that she was determined?
2. How did Dr. Browne show her generosity?

! explore it

Dr. Browne went to school during the Great Depression of the 1930s. Find out more about the Great Depression. Why was this period a particularly difficult one for many African Americans?

Marjorie Lee **Browne**
TEACHING NOTES

Purpose: To introduce students to one of the nation's first African-American women to earn a Ph.D. in mathematics.

Mathematics Background: Marjorie Lee Browne received her Ph.D. from the University of Michigan in 1949. Her doctoral dissertation was entitled "The One Parameter Subgroups in Certain Topological and Matrix Groups." A one parameter group is a finite or infinite set of elements or spaces with only one variable. A topological group has no holes; a simple example is the set of real numbers. Dr. Brown published a paper, "A Note on Classical Groups" in the *American Mathematical Monthly* in 1955, and she researched computing and numerical analysis at the University of California at Los Angeles. She continued her research of topological groups at Cambridge University and Columbia University.

1. INTRODUCE

About Marjorie Lee Browne: Browne earned her bachelor's degree from Howard University in 1935 graduating cum laude. In 1939 she received her master's degree in mathematics from the University of Michigan. Ten years later, she earned her Ph.D. in mathematics.

About Browne's Work: Dr. Browne taught mathematics during much of her career. She had stints at the Gilbert Academy in New Orleans, Wiley College in Marshall, Texas, the University of Michigan, and North Carolina Central University, where she became chair of the Mathematics Department. She also had fellowships at Cambridge University, the University of California at Los Angeles, and Columbia University. Dr. Brown was the first recipient of the W. W. Rankin Memorial Award for Excellence in Mathematics Education. The award is given by the North Carolina Council of Teachers of Mathematics.

2. DISCUSS

Use the copy master to prompt discussion. Talk with students about Dr. Browne's generosity. Help them understand that, by virtue of her family's relatively good fortune during the Depression, she probably felt fortunate to have been able to have the opportunity to get a good education and develop a successful career. At the end of her career she wanted to give other minority students the same chances she had.

Answers to Explain It:

1. Answers will vary. Possible answer: Browne went to school part-time for nearly 10 years to get her Ph.D.

2. Answers will vary. Possible answer: Browne gave money out of her own pocket to minority students who wanted to get math educations.

3. EXTEND

After students have completed the Explore It, ask them to share what they have learned. Help students understand that African Americans generally did not fare very well during the Great Depression. Lynchings and mob violence against them increased during this period.

Read stories about the Depression that were told by children from that period.

Career Focus: Mathematicians use a number of important tools in their work. What do you think is their most important tool?

Jewell Plummer Cobb

AT A GLANCE

born 1924
Chicago, Illinois

education
B.S., Talledega College
Ph.D., New York University

occupation
Biologist, Cell Physiologist

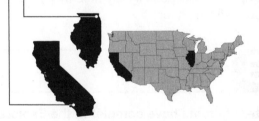

Jewell Plummer Cobb has a great challenge. She wants to find a treatment for *melanoma*. This is the most serious form of skin cancer. About 7,600 people in the United States died from this disease in 2003. The number of cases is rising, too.

Your skin is damaged by spending too much time in the sun without sunscreen. This damage can result in skin cancer years later. A person's family history also plays a role. Cancer occurs when cells grow and divide much faster than normal. As they grow, the cancer cells destroy healthy cells.

Dr. Cobb is an expert in cells. She has a degree in *cell physiology*, the study of how cells work. Her studies have resulted in new drugs to treat skin cancer. There is still much work to do, however.

Using sunscreen helps prevent skin cancer.

Besides doing research, Dr. Cobb directed the work of several laboratories. She also served as president of California State University at Fullerton. While there, she helped make it possible for more minorities to attend the school.

Although Dr. Cobb is retired, she still devotes much of her time to research. She wants to help people with skin cancer. Dr. Cobb also has started several programs that help minority students and women pursue careers in science.

? explain it

1. What kinds of things does a cell physiologist study?
2. Why would a cell physiologist be interested in cancer?
3. What do you think is important about Dr. Cobb's work?

! explore it

Much information about the prevention of melanoma exists. Locate some of this information. Then create a leaflet that highlights the main things people can do to avoid developing melanoma or other skin cancers.

Jewell Plummer Cobb
TEACHING NOTES

Purpose: To introduce students to a scientist who does research on cancer and encourages minorities and women to enter scientific fields.

Science Background: Today, more than 200 types of cancers have been identified. Most have been identified by two means: discovery of the site where the cancer originates and the appearance of the cancerous tissue when viewed under a microscope. Each kind of cancer is named for the tissue in which it develops. For example, *sarcoma* is a term that denotes any cancer that originates in connective tissue. *Melanoma* is a term that denotes any malignant cancer that originates in the melanocytes, the cells containing color.

What causes cancer? Scientists do not know for certain. However, they do know that exposure to particular substances can increase a person's chances of developing cancer. These substances are called *carcinogens* and are associated with between 60 and 90 percent of all human cancers. Another cause of cancer is viruses. More than 100 separate viruses have been associated with types of cancer. Most of these viruses have been found in the tissues of other species of animals, but not in humans. Stress may also play a role. Given the same genetic and environmental factors, it has been proposed that people who are under a great degree of stress are more likely to develop cancer than others.

I. INTRODUCE

About Jewell Plummer Cobb: Dr. Cobb is a strong, positive role model. Her dedication to research, as well as her commitment to helping minorities and women obtain better opportunities in higher education, provides an excellent opportunity to discuss habits of mind. Ask students how Dr. Cobb's enthusiasm has affected her success.

About Cobb's Work: Dr. Cobb has published more than 35 articles concerning her cancer research. She has been awarded grants for her work by such diverse organizations as the National Institutes of Health, the Laboratory Internazionale di Genetica e Biofisica, and the National Science Foundation. Awards for her achievements are numerous. They include many honorary degrees from prestigious universities.

2. DISCUSS

Use the copy master to prompt discussion. Lead students to understand that scientists and physicians know what the result of cancer is but do not know yet what causes it. Make sure that students realize that each type of cancer is different and that some forms can be treated more easily than others. In addition, encourage students to realize that only through continuing research can we learn enough about cancer to control, cure, or eliminate it.

Answers to Explain It:

1. Cell physiologists study how cells work.

2. Answers will vary. Possible answer: A cell physiologist would be interested in cancer because cancer occurs when cells grow and divide much faster than normal.

3. Answers will vary. Possible answer: Her work is important because it may help many people recover from cancer.

3. EXTEND

After students have completed the Explore It, ask them to share what they have learned. Students' leaflets should include suggestions such as wearing sunglasses, avoiding the sun between 10 a.m. and 3 p.m., using a sunscreen with an SPF no lower than 15, wearing a hat, and so on.

Career Focus: Discuss with students some of the things that they can do now to prepare for a career in cell physiology. What courses should you take? What magazines should you read? Who could you talk with to get information?

Marie Maynard Daly

AT A GLANCE

lived 1921–2003
Corona, New York

education
B.S., Queens College
M.S., New York University
Ph.D., Columbia University

occupation
Chemist, Professor

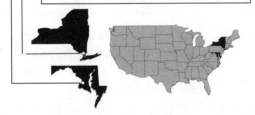

Marie Maynard Daly came from a family who loved science. Her grandfather had a huge library filled with many science books. Daly's mother often read these books to her. Her father went to school to become a chemist. But he was unable to complete his education because he could not afford to get his degree.

With such an upbringing, it was only natural for Daly to want to be a scientist. To help reach her goal, she went to Hunter College High School. This school was known for encouraging African-American women to train for careers.

○ **DALY** researched the aging process.

The teachers at Hunter College High School trained her well. Daly went on to earn two degrees in chemistry. Then, in 1947, she became the first African-American woman to earn a Ph.D. in chemistry.

As a chemist, Dr. Daly researched the process of aging and studied the reasons why some people get high blood pressure. She also taught chemistry at several universities.

Dr. Daly retired in 1986. In 1988 she started a fund for African-American students in memory of her father. This fund has helped many students with their science educations.

? explain it

1. What role did Daly's family play in her education?
2. Why do you think it is important to study high blood pressure?

! explore it

People with a background in chemistry work in many different fields. Find out about some of them. Create a chart that shows some of the jobs a chemist can do.

Marie Maynard **Daly**
TEACHING NOTES

Purpose: To introduce students to the first African-American woman to earn a Ph.D. in chemistry.

Science Background: Researchers have long known about the factors that damage blood vessels—smoking, hypertension, lack of exercise, and cholesterol. Yet they still do not know why some people are able to repair damage caused by these factors while others are not.

Researchers at Duke University Medical Center believe they know the answer. According to Dr. Pascal Goldschmidt, a cardiologist, "We believe that the key resides in the bone marrow, which produces cells that can repair damage to the body." Some people produce more of these cells while others produce fewer. Those who produce more remain healthier longer than others. Ultimately the Duke researchers believe that it might be possible to forestall or even prevent atherosclerosis by injecting stem cells into the bodies of people who are deficient in these beneficial cells. It might even be possible to "retrain" a person's own stem cells to become capable of repairing arterial damage.

1. INTRODUCE

About Marie Maynard Daly: Daly graduated magna cum laude from Queens College in Flushing, New York, in 1942. She earned a bachelor's degree in chemistry, and a year later she earned her master's degree from New York University. In 1947 Columbia University awarded her with a Ph.D. in chemistry. She was the first African-American woman to earn a doctorate in chemistry.

About Daly's Work: Dr. Daly taught chemistry at Howard University and Columbia University. She also was an associate professor at the Albert Einstein College of Medicine at Yeshiva University. Her research included the study of the metabolism of arterial walls and how this process is related to aging, hypertension, and atherosclerosis. She also studied nucleic acids and the composition and metabolism of components of cell nuclei.

2. DISCUSS

Use the copy master to prompt discussion. Lead students to conclude that Daly's family played an integral role in her education and her eventual professional success.

Discuss with students some of the century's most important medical breakthroughs. Help students conclude that new breakthroughs might also be made in the areas of high blood pressure and aging. These breakthroughs could help extend people's lives or improve their quality of life.

Answers to Explain It:

1. Answers will vary. Possible answer: Her family supported her. Her mother read her books from her grandfather's library. Her father was also a good influence because he had an interest in chemistry.

2. Answers will vary. Possible answer: New medicines or treatments might be discovered that will help extend the lives of many people.

3. EXTEND

After students have completed the Explore It, ask them to share what they have learned. Discuss some of the traits that a person desiring a career in chemistry should have.

Career Focus: Think about some of the major health problems that people today have. If you were a scientist, which one of these health problems would you like to investigate? What might be the result of your work?

Christine Mann Darden

AT A GLANCE

born 1942
Monroe, North Carolina

education
Hampton University
M.A., Virginia State University
Ph.D., George Washington University

occupation
Mechanical Engineer,
NASA Researcher

Did you know a sonic boom can be dangerous? Dr. Christine Mann Darden is an expert in sonic boom technology. She works for NASA's Langley Research Center in Hampton, Virginia, and her job is to study sonic booms.

Sonic booms are loud sounds similar to explosions or thunder. They are caused by aircraft that go faster than the speed of sound. Large shock waves are created, which cause the loud sounds when they reach Earth. These waves are so strong that they sometimes break windows. They can also damage other structures.

Dr. Darden studies sonic booms for several reasons. One reason is to be able to predict how a sonic boom might affect a particular area. This is done with experimental

DARDEN tests sonic boom technology.

airplanes. Another reason is to find new ways to reduce their damage.

Dr. Darden has worked for NASA since 1967. Before that she earned advanced degrees in math and engineering. She credits her parents with teaching her about the importance of a good education.

Dr. Darden received the Women in Engineering Lifetime Achievement Award. She has been a strong supporter of increasing the number of women and minorities in the sciences.

? explain it

1. Why can sonic booms be considered a problem?
2. What might be a major result of the work done by Darden?

! explore it

Sonic booms are created by aircraft that break the speed of sound. Find out what the speed of sound is. How long have planes been capable of breaking the sound barrier?

Christine Mann
Darden
TEACHING NOTES

Purpose: To introduce students to a successful mechanical engineer and NASA researcher.

Science Background: The speed of sound at sea level is approximately 1225 kilometers per hour (761 mph). As altitude increases, the speed of sound decreases. For example, at 20,000 feet an aircraft would have to travel 660 miles per hour to reach the speed of sound.

Speeds beyond the speed of sound are measured using Mach (M) numbers. This measurement is named after its developer, Austrian physicist Ernst Mach. Using this system the speed of sound is Mach 1, twice the speed of sound is Mach 2, and so on.

The first aircraft to break the sound barrier was the X-1, flown by famed pilot Chuck Yeager in 1947. Mach 3.3 was achieved by the SR-71 in the 1960s. The current speed record is Mach 9.8. NASA's X-43A Scramjet traveled at nearly 10 times the speed of sound (7,000 mph) on November 16, 2004. Scientists hope a plane like the Scramjet will one day take astronauts into space.

1. INTRODUCE

About Christine Mann Darden: Christine Mann Darden is the director of the Aero Performing Center Program Management Office at the NASA Langley Research Center in Hampton, Virginia. She holds a master's degree in applied mathematics from Virginia State University and a Ph.D. in mechanical engineering from George Washington University.

About Darden's Work: Dr. Darden started her NASA career as a data analyst in 1967. In 1972

she accepted an engineering position, subsequently researching sonic boom technology. Over the years she has become an expert in speeds greater than sound. She is the author of more than 50 articles and technical papers dealing primarily with sonic boom prediction, sonic boom minimization, and supersonic wing design.

2. DISCUSS

Use the copy master to prompt discussion. Students should realize that sonic booms can be powerful enough to break windows and cause structural damage to some buildings. The problems associated with sonic booms have limited the expansion of supersonic travel in the United States.

Lead students to understand that Darden's work might eventually lead to new designs for supersonic aircraft. These designs will greatly minimize or eliminate the damage caused by sonic booms.

Answers to Explain It:
1. Answers will vary. Possible answer: Sonic booms are a form of noise pollution. They could be particularly bothersome for elderly or sick people. Sonic booms can also damage buildings and break windows.

2. Answers will vary. Possible answer: New aircraft might be developed that do not cause as much damage as they do now.

3. EXTEND

After students have completed the Explore It, ask them to share what they have learned. Share with students the information presented in Science Background.

Career Focus: Imagine that you have been given the opportunity to work for NASA. What type of work would you like to do there? What preparations and education do think you would have to have?

Mark Dean

AT A GLANCE

- **born** 1957
  Jefferson City, Tennessee
- **education**
  B.S., University of Tennessee
  Ph.D., Stanford University
- **occupation**
  Computer scientist, Inventor

How did you last use a computer? Maybe you wrote a paper or played a game. Dr. Mark Dean is one of the people who made personal computers possible. In the 1980s, Dr. Dean went to work for IBM. His group created the original IBM personal computer.

Dr. Dean became interested in building things when he helped his dad build a tractor from scratch. Dean became an athlete and a straight-A student in high school. Today, he encourages minority students to pursue science and engineering. He enjoys working on the "human part of technology." Dean focuses on how computer science can best serve people. He thinks this focus will encourage more young people to join the field.

○ **DEAN** invented a system that gave us PCs.

As head of an IBM research center, Dr. Dean oversees about 400 scientists and engineers. But he does not believe in all work and no play. One recent spare-time project was making a replica of a classic hot rod. He installed a video camera and a laptop so he could record his cross-country adventures. But work itself may be the most fun for Dr. Dean. When asked about his work at IBM, he says, "I'm having fun the whole time."

? explain it

1. Dr. Dean thinks computer science should focus on people instead of just technology. Why?

2. What characteristics of a job might make work fun for employees?

! explore it

What is a mainframe computer? How does it differ from a personal computer? Why was the personal computer such an important invention?

Mark Dean

TEACHING NOTES

Purpose: To introduce students to a computer scientist whose inventions helped make the personal computer possible.

Science Background: Mark Dean was part of a team at IBM in the early 1980s in charge of creating a personal computer. According to Dean, IBM gave its personal computer team much independence: "It was a really fun time.... It was a pretty basic design...a tool that was essentially better than a typewriter and a calculator and a hand spreadsheet—all those other little pieces that you had to have when you sat down to do your business."

With Dennis Moeller, Dean developed the "ISA systems bus." This key invention made it possible for a PC to be connected to a variety of devices such as a printer and a modem. Later, Dean led a team that built a chip enabling a computer to make 1 billion calculations per second. Projects that interest Dean today include moving beyond hard drives and replacing the current semiconductor chips with more effective technology. He is also interested in technology that would enable individuals to automatically record just about every aspect of their private and business lives.

About Dean's Work: At IBM, Dean works to recruit scientists and engineers from minority groups. He says, "A lot of kids growing up today aren't told that you can be whatever you want to be. There may be obstacles, but there are no limits." Dean was elected to the National Inventors Hall of Fame for inventing "a system that has allowed PCs to become part of our lives." He was named one of the 50 Most Important Blacks in Research Science. About careers, Dean says, "First, have fun, because you put too many hours in not to enjoy your work; and second, nothing is more important than family—so be flexible with your time."

2. DISCUSS

Use the copy master to prompt discussion. Have students discuss possible reasons for Dean's success based on the details about his career and personal traits.

Answers to Explain It:
1. Answers will vary. Possible answer: He thinks this will encourage students to become computer scientists. Students might be more excited about training for a career working with and serving people instead of one based solely on working with technology.

2. Answers will vary. Possible answer: A fun job might be one that is creative, offers some independence, and takes full advantage of one's interests and talents.

I. INTRODUCE

About Mark Dean: Dean's mother was a teacher. His father was a supervisor of Tennessee Valley Authority dams. Dean studied trigonometry and geometry in first grade. He was an excellent student and athlete in high school, but also played drums in a band. Dean says that his parents had to struggle against discrimination in ways that he has not. He says, "They have accomplished more than I have." One of his University of Tennessee professors called Dean "the best student I ever had." Dean went to Stanford for his Ph.D. after already accomplishing goals at IBM.

3. EXTEND

After students have completed the Explore It, ask them to share what they have learned. A mainframe is a large, powerful, expensive computer used by many people at once. A personal computer, on the other hand, is small, portable, and affordable. It is meant to be used by one person at a time. The personal computer provides computer access to people who could never use mainframe computers.

Career Focus: Computer science researchers like Dr. Dean look for new uses for computer technology. What skills do you need to become a computer scientist?

Nathaniel **Dean**

AT A GLANCE

born 1956
Mound Bayou, Mississippi

education
B.S., Mississippi State University
M.S., Northeastern University
Ph.D., Vanderbilt University

occupation
Mathematician, Professor

Some people think they have no need to learn math. They couldn't be more wrong. Math affects our lives in many ways. It is a driving force in the computer age in which we live.

Nathaniel Dean learned early in life how math affects our lives. That's why he made it a goal to get a good math education. He earned degrees at Mississippi State and Northeastern universities. Then he received a Ph.D. from Vanderbilt University.

Dr. Dean is an expert in graph theory. This field of math can be used to solve many kinds of everyday problems. For example, a graph can be used to show how one computer is connected to other computers. It is important to study these links in order to know what to do if something goes wrong.

DR. DEAN is the chair of the Mathematical Sciences Research Institute.

Dr. Dean has had many jobs during his career. He has designed computer software for a large company and has taught math to college students. Dr. Dean has also written books and articles about mathematics.

Math is not the only thing that interests Dr. Dean. In addition to enjoying the martial arts, he helps minority students and supports youth leagues. Dr. Dean actively encourages young people to get good educations.

? explain it

1. Why is it important to study math?
2. Could you say that Dr. Dean's career has been varied? Explain your answer.

! explore it

Nathaniel Dean designed software for a large company after he graduated from Vanderbilt University. What is software? What kinds of software might be useful to a student?

Nathaniel Dean
TEACHING NOTES

University, Brown College, and Texas Southern University, where he is dean and chair of the Mathematical Sciences Research Institute. He has written several books and numerous papers dealing with mathematics and computer science. His research emphasis is graph theory.

Purpose: To introduce students to a successful mathematician who is an expert in graph theory.

Mathematics Background: Dean is considered an expert in graph theory, the study of relations on finite sets that can be visualized with dots and lines. Graphs serve as models for many problems in science, business, and industry. Theorems about graphs can be used to solve problems involving scheduling, networking, and routing. An example would be to make a graph that shows visually how a particular desktop computer is connected to other computers. Such a graph can be particularly useful for predicting vulnerability in the network. It can show what would happen to all of the other computers if one of them experienced a problem, such as being disconnected from the Internet. Graph theory has many applications in other fields. For example, in mathematical chemistry, molecules are modeled by graphs. Graph theory is the basic tool of research in the theory of communications networks. It is also useful in linguistics and genetics.

1. INTRODUCE

About Nathaniel Dean: Dr. Dean is an active member in minority organizations and youth leagues. He received his bachelor's degree in mathematics from Mississippi State University, his master's degree in applied mathematics from Northeastern University, and his Ph.D. from Vanderbilt University. His doctoral thesis is entitled "Contractible Edges and Conjectures About Path and Cycle Numbers."

About Dean's Work: Dr. Dean worked in the Software Production Research Department at Bell Laboratories between 1987 and 1998. Since then he has held teaching positions at Rice

2. DISCUSS

Use the copy master to prompt discussion. Lead students to understand that mathematics and computers are inexorably tied. Discuss some of the devices that students use that contain computer chips. Talk about the use of mathematics related to the programming of these devices.

Answers to Explain It:
1. Answers will vary. Possible answer: It can be used to solve all sorts of everyday problems, particularly those that present themselves in computer networks.

2. Answers will vary. Possible answer: Dean has had a varied career. He has written books and articles, taught math to college students, and worked on developing software programs. In addition, he is involved in community work.

3. EXTEND

After students have completed the Explore It, ask them to share what they have learned. Help students understand that computer software contains the instructions for a computer or computer network. Examples of software useful to most students: word programs and photographic programs. Ask students what types of software might be useful to a math student. A program that helps create graphs or charts would be useful.

Career Focus: Many computer-related jobs require a strong background in mathematics. What other type of training or background do you think you would need to work with computers?

Charles Drew

AT A GLANCE

lived 1904–1950
Washington, D.C.

education
B.A., Amherst College
M.D., McGill University
DSc (Med), Columbia University

occupation
Surgeon, Researcher, Professor

Before World War II (1941–1945), sick and injured people often died because they could not get blood. No one had yet thought of a way to store it for long periods. In 1939, Dr. Charles Drew developed a way to save and store blood plasma. Plasma is the clear, sometimes yellowish, part of blood. Blood cells float in this liquid.

Dr. Drew's discovery helped save many lives during World War II. At that time, he was asked to work on the Blood for Britain project. For this project, Dr. Drew and others created a collection of dried blood plasma that was sent to injured soldiers in Britain.

After the success of the Blood for Britain project, Dr. Drew became the first director of the Red Cross Blood Bank. He started the use of "blood-mobiles," refrigerated trucks to carry donated blood. Then, in 1941, he became a professor of surgery at Howard University Medical School.

Blood bank trucks used to carry donated blood

Today, hospitals and blood banks safely store blood for use by sick and injured people. Many people help by donating blood.

Dr. Drew's work on stored blood has saved lives. Yet it failed to help him. In 1950 he was injured in a car crash and needed blood. The closest hospital took only whites. He died enroute to the African-American hospital.

? explain it

1. Why is it important to be able to save and store blood?
2. How does the work of Dr. Drew affect you today?

! explore it

Research the blood banks in your community. Find the answers to the following questions: How often do the blood banks hold blood drives? What can you do to help recruit donors during a blood drive?

Charles **Drew**
TEACHING NOTES

Purpose: To introduce students to a doctor who was instrumental in developing a method for storing blood plasma for transfusions.

Science Background: Blood can be grouped based on antigens present in the red blood cells. The two major classifications are the ABO grouping and the Rh system. The ABO blood grouping is based on two antigens: A and B. In turn, the presence of these antigens makes up the four blood types. A person who has only A antigens has blood *type A*. A person who has only B antigens has blood *type B*. A person who has both A and B antigens has blood *type AB*. A person who has neither A nor B antigens has blood *type O*.

The blood plasma of many people has antibodies that act against any blood type other than their own. For that reason, when transfusions are given, it is vital to provide the person with his or her own blood type.

The Rh type of the blood is also important in transfusions. People who have antigens for the Rh factor are said to be Rh+ (Rh positive). Those without the antigens are Rh– (Rh negative). Under normal circumstances, human plasma does not contain anti-Rh antibodies. However, if a person has been exposed to blood of another Rh type, he or she will form those antibodies. Once the antibodies are formed, a transfusion of the incorrect Rh type may result in the person's death.

I. INTRODUCE

About Dr. Charles Drew: Before becoming a doctor, Charles Drew was most well-known for his prowess on the athletic field. At Dunbar High School, he received an award for all-around athletic performance. At Amherst College, he was the quarterback of the football team, the captain of the track team, and was the most valuable player on his baseball team.

Drew was also a humanitarian. His work was recognized worldwide. He received honorary degrees from Virginia State College and from Amherst. In 1944, he received the Spingarn Medal from the National Association for the Advancement of Colored People.

About Drew's Work: During World War II, the U.S. military decided to segregate plasma supplies for white and black soldiers. It was believed that this should be done because of the possibility that different races of humans had different types of blood. Upset with this directive, Drew promptly conducted studies that showed conclusively that all humans have the same blood types, regardless of race.

Dr. Drew became a victim of segregation in 1950. He was wounded in a serious car accident, and the ambulance was turned away from the closest hospital, a whites-only building. He died in the ambulance on the way to a hospital for African Americans.

2. DISCUSS

Use the copy master to prompt discussion. Ask students how having blood ready for transfusion has changed the lives of people.

Answers to Explain It:
1. It is important to have blood available for emergencies.

2. Answers will vary. Possible answer: Due to his work, if I ever need a blood transfusion I will probably be able to get one. It might save my life.

3. EXTEND

After students have completed Explore It, ask them to share what they have learned. Interested students might like to create posters designed to encourage people to become donors at one of the blood drives in their community.

Career Focus: You have been working in a hospital as a student intern for several weeks. The chief of staff is impressed with your job performance and wants you to consider becoming a doctor. What might be your response? Why?

Euphemia Lofton Haynes

AT A GLANCE

lived 1890–1980
Washington, D.C.

education
B.A., Smith College
M.A., University of Chicago
Ph.D., Catholic University

occupation
Mathematics and English teacher

Some people like to teach. Others like to learn. Dr. Euphemia Lofton Haynes had a great love for both.

Dr. Haynes was born and raised in Washington, D.C. Her father was a prominent dentist and her mother was active in the Catholic Church. Haynes graduated with honors from high school in 1909. She received a bachelor's degree in math and began teaching math and English to students in Washington, D.C. Dr. Haynes would go on to teach for 47 years in the city's public schools.

Dr. Haynes continued learning while she was teaching. After she received a master's degree in education, she became a student at Catholic University. In 1943 the school awarded her a doctorate degree in mathematics. Dr. Haynes became the first African-American woman to attain such a degree. She was 53 years old at the time.

○ **HAYNES** taught children in Washington, D.C.

Dr. Haynes did much more than teach and learn. She loved to help people in her community. She was the head of the Washington, D.C., school board. In addition, she helped African-American students gain many rights. One of these was the right to enter the schools of their choice.

? explain it

1. What do you think was Dr. Haynes' most important accomplishment? Why?
2. In what ways did Dr. Haynes help the people in her community?

! explore it

Explore the Internet to find information about a math-related career. Prepare a fact sheet with information about type of degree or training needed, description of the job, and salary.

Euphemia Lofton Haynes
TEACHING NOTES

Purpose: To introduce students to a mathematician who had a long and successful career as a teacher and community activist.

Mathematics Background: Young women of today are encouraged to seek careers in math and science. This was not always the case. During the years when Dr. Euphemia Lofton Haynes attended school, there were very few opportunities and many social roadblocks put in front of women who aspired to be mathematicians or scientists, especially minorities. Despite the obstacles, Haynes carved a successful career for herself through her determination and drive. Throughout history, women with similar drives also overcame social barriers to become successful mathematicians. Here are two of them:

Emilie du Chatelet, France (1706–1749), famous for translating Newton's *Principia* from Latin to French, had to disguise herself as a man in order to study math and science.

Christine Ladd-Franklin, U.S. (1847–1930), an expert in symbolic logic, was denied a Ph.D. by Johns Hopkins University because it did not grant advanced degrees to women. She finally got her degree, 44 years later, and became a lecturer at Johns Hopkins as well as Columbia University.

1. INTRODUCE

About Euphemia Lofton Haynes: Dr. Haynes was dedicated to the pursuit of knowledge and the teaching of that knowledge. She had lofty educational goals that were achieved after many years of study. Share with students that Haynes did not receive her doctorate degree until she was 53. Ask students what this fact tells them about her personality.

About Haynes' Work: Dr. Haynes became the first African-American woman to receive a doctorate in mathematics. She was also the first woman to chair the Washington, D.C., school board. Dr. Haynes taught English and math in the Washington, D.C., public school system until her retirement in 1959. She helped establish the Mathematics Department at Miners Teachers College and occasionally taught at Howard University. In addition, she actively supported numerous community projects. Dr. Haynes was a member of the American Association of University Women, League of Women Voters, NAACP, Catholic Interracial Council, National Conference of Christians and Jews, and many other organizations. She also was involved in the integration of the Washington, D.C., public school system.

2. DISCUSS

Use the copy master to prompt discussion. Help students understand that community activists such as Dr. Haynes come from a variety of backgrounds and professions. Discuss some of the attributes a community activist should have.

Answers to Explain It:
1. Answers will vary. Possible answer: Her most important accomplishment was the education of all the students she taught during her 47 years in the classroom. She gave these students the background necessary for developing their own careers.

2. She helped make it possible for African-American students to attend the schools of their choice.

3. EXTEND

After students have completed the Explore It, ask them to share what they have learned. Mathematicians have an opportunity to make a lasting contribution to society by helping to solve problems in such diverse fields as medicine, management, economics, government, computer science, physics, psychology, engineering, and social science.

Career Focus: Teachers work long and hard hours, and they often continue their educations while they are teaching. Does this type of job appeal to you? Why or why not? What traits do you think you would need to be a good teacher?

Katherine G. Johnson

AT A GLANCE

born 1918
White Sulphur Springs, West Virginia

education
B.A., West Virginia State College

occupation
Physicist, Space Scientist,
Mathematician

Katherine Johnson was a pioneer. In the 1950s she began working as a scientist for the U.S. space program. Few women at the time had such jobs.

Johnson grew up on a small farm in West Virginia. During most winters, the roads in the mountains would get blocked with snow. Johnson's father moved the family to town so that his children could go to school during the winter.

Johnson majored in mathematics and French in college. She then taught school for several years.

In 1953 Johnson began working for NASA, an agency that runs the U.S. space program. Johnson worked on the Apollo moon missions in

○ **JOHNSON** worked on the Apollo moon missions.

the 1960s and early 1970s. She studied data about the spacecraft as they orbited the moon. Later she helped determine ways to guide the spacecraft back to Earth.

Johnson also worked with the Earth Resources Satellite. This satellite was used to find minerals and other resources within Earth.

NASA gave Johnson an award for her work on the Apollo program. She retired after working for NASA for more than 30 years.

? **explain it**

1. How did Johnson's family help her become a successful scientist?
2. Why do we test spacecraft (without pilots) before astronauts fly in them?
3. Why do you think the Earth Resources Satellite was important?

! **explore it**

Find out about the "space race" that began in the 1950s. What caused the United States to develop its space program? How did the beginning of the space program affect students?

Katherine G. Johnson
TEACHING NOTES

tions during the Apollo lunar missions. In addition, she calculated the trajectories needed to obtain orbit around the moon and land on the surface. She was a recipient of the Group Achievement Award that was presented to NASA's Lunar Spacecraft and Operations team.

Purpose: To introduce students to a physicist who did important work for NASA during the Apollo missions.

Science Background: On October 4, 1957, the Soviet Union launched an artificial satellite into space. The satellite, Sputnik 1, was the size of a basketball and carried scientific instruments to measure the density and temperature of Earth's upper atmosphere. It orbited the Earth in about 98 minutes. Sputnik II was launched in November and included the first organism in space, a dog named Laika. The launch of the first U.S. satellite followed closely, in January 1958. Explorer 1 carried scientific equipment to measure temperature and cosmic rays.

Although the United States was not far behind the Soviet Union, the Soviet success caused panic in the Eisenhower administration. New curricula stressing math and science were instituted in schools throughout the country. Children were tested and "tracked" to become the scientists the United States would need to compete with the Soviet "machine." The "space race" was on. As a result, Congress passed the Space Act in 1958, which created the National Aeronautics and Space Administration (NASA).

I. INTRODUCE

About Katherine Johnson: Johnson is a talented, determined woman who overcame many obstacles to have a successful career at NASA. She created a successful career despite the racial and gender barriers in place at the time. Ask students: How does Johnson serve as a role model in her work as a scientist? What could you do if you wished to follow in her footsteps?

About Johnson's Work: Johnson worked on the problems associated with trajectories between planets, and navigating spacecraft in and out of orbit. She analyzed data gathered at tracking sta-

2. DISCUSS

Use the copy master to prompt discussion. Lead students to understand that when the first spacecraft were flown, little was known about launching, orbiting Earth, or landing. Using craft that had no pilots enabled technicians to be prepared when a human was sent into space.

Answers to Explain It:
1. Johnson's father moved the family from the mountains to the town of White Sulphur Springs. This move ensured that the Johnson children would not miss school during the winter. Part of Johnson's success in later years could be attributed to the early actions of her father.

2. Answers will vary. Possible answer: It enables scientists the opportunity to discover problems with various instruments and systems.

3. Answers will vary. Possible answer: It was important because it enabled us to locate important minerals that conventional techniques might have missed.

3. EXTEND

After students have completed the Explore It, ask them to share what they have learned. Students may wish to gather information from older family members. Personal memories of family members may prove just as or more enlightening than information read in a book. However, you may wish to supplement students' contributions with the information found in the Science Background on this page.

Career Focus: Johnson is a physicist and mathematician. What other kinds of jobs involve using physics or mathematics?

Theodore K. Lawless

AT A GLANCE

lived 1892–1971
Thibodeaux, Louisiana

education
Talladega College, University of
Kansas, Columbia University,
Harvard University,
M.D., Northwestern University

occupation
Dermatologist

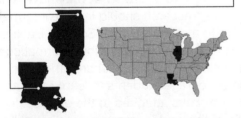

Dr. Lawless grew up in a small town in Louisiana. But his studies took him to places around the world. He earned his medical degree from Northwestern University in Evanston, Illinois. He then studied and taught in Europe.

Dr. Lawless specialized in *dermatology*, a field that deals with the skin and its diseases. During his career, he helped develop medicines and treatments for people with skin diseases.

He loved his work. Dr. Lawless saw about 100 patients a day in his Chicago office. Why did he work so hard? He once said, "I am happiest when I work, and so I work as much as I can."

His practice kept him busy. Yet he still found time to teach at Northwestern University.

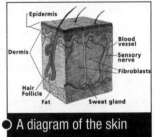

A diagram of the skin

Dr. Lawless received many honors. He won the Harmon Award for outstanding achievement in medicine. He also received the Spingarn Medal. This award is given each year to one outstanding African American.

Dr. Lawless was a millionaire when he died in 1971. He gave most of his money away. He gave a large gift to Dillard University in New Orleans and to a hospital in Israel. The hospital used the money to build a skin clinic that bears his name.

? explain it

1. What was Dr. Lawless' attitude toward his work?
2. Why is it important to like your work?
3. What evidence suggests that Dr. Lawless was a generous man?

! explore it

Much of Dr. Lawless' work centered around the disease of leprosy. Find out about this disease. How does it affect the skin? What treatments are available? How common is it today?

Theodore K. Lawless

TEACHING NOTES

Purpose: To introduce students to a dermatologist who developed medicines and treatments for several severe skin diseases.

Science Background: Dr. Lawless made significant contributions toward the understanding and treatment of leprosy.

Leprosy, also known as Hansen's disease, is a chronic, infectious disease caused by the rod-shaped bacillus *Mycobacterium leprae*. The disease affects many parts of the body, particularly the eyes, skin, fingers, toes, and mucous membranes of the upper respiratory tract.

Somewhat surprisingly, leprosy has not been eradicated from the world. In fact, in 2003 health officials estimated that there were about 1.5 million cases in the world. Most of these were concentrated in Southeast Asia, Africa, and Latin America.

The first effective treatment for leprosy, dapsone, became available in the late 1940s. Through the years, however, bacilli resistant to the drug began appearing. Today, multidrug therapy (MDT—a combination of three antibiotics) has proven very effective in killing the germs. Since its introduction in the early 1980s, the drug regimen has cured more than 8 million people afflicted by the disease.

I. INTRODUCE

About Theodore K. Lawless: Dr. Lawless earned his medical degree from Northwestern University in 1919. He taught dermatology at Northwestern from 1924 to 1941. He also was the senior attending physician at Provident Hospital. Dr. Lawless had a thriving private practice located in the heart of Chicago's African-American community. Through the years, he saw an average of 100 patients a day.

About Lawless' Work: Dr. Lawless developed treatments for sexually transmitted diseases, including medication for the side effects

associated with the treatment of syphilis. He also developed treatments for Hansen's disease and experimented with the use of radium in the treatment of cancer.

2. DISCUSS

Use the copy master to prompt discussion. Lead students in a discussion of Dr. Lawless' work, particularly his positive attitude and his work ethic. Then discuss why such an attitude is helpful in every job or career, regardless of how much it pays monetarily.

Answers to Explain It:

1. Answers will vary. Possible answer: Dr. Lawless greatly enjoyed his work. He said, "I am happiest when I work, and so I work as much as I can."

2. Answers will vary. Possible answer: When you like your work, it is not a chore; it is an enjoyable experience that you can look forward to every day.

3. Answers will vary. Possible answer: He donated money to colleges and hospitals and other causes.

3. EXTEND

After students have completed the Explore It, ask them to share what they have learned. Students' research should include information similar to that found in the Science Background notes on this page. Share this information with students after they have shared their findings. Also encourage students to learn more about the social stigmas attached to the disease. In some parts of the world leprosy victims are still segregated in separate colonies.

Career Focus: Dermatologists treat a wide variety of skin diseases and conditions. What would you consider to be some of the negative and positive aspects of working as a dermatologist?

E. Don Sarreals

AT A GLANCE

born 1931
Winston-Salem, North Carolina

education
B.S., M.S., New York University

occupation
Meteorologist

Predicting tornadoes and other types of weather is the science of meteorology. E. Don Sarreals had a successful career as a meteorologist. He is now retired. Sarreals began his career as a forecaster for the National Weather Service (NWS). Besides forecasting the weather, he studied the atmosphere, winds, temperature, and moisture.

The weather changes every day. Some changes can be dramatic. For example, a tornado might sweep into a town and cause great damage. It might also injure or kill people.

Knowing when a tornado is likely to strike can save lives. Sarreals used special tools to help him predict the weather. One of these tools is radar. Radar helped

NEXRAD measures both rain and wind to predict the weather.

Sarreals see how the air in the atmosphere was moving. Tight swirls of air might mean a hurricane is developing. Heavy clouds can indicate a thunderstorm.

Satellites are used to send the signal for the radar images to Earth. Weather changes can be seen as they happen.

For Sarreals' last job, he was the head of the NEXRAD project, short for "next generation radar." This type of radar accurately measures both rain and wind.

? explain it

1. How are weather forecasters helpful to you?
2. How do meteorologists use radar to help forecast the weather?
3. What part do satellites play in weather forecasting?

! explore it

The NWS is responsible for warning people about severe weather. Find out what types of severe weather bulletins the NWS issues. Make a poster that shows how to stay safe in severe weather.

E. Don **Sarreals**
TEACHING NOTES

Purpose: To introduce students to a scientist who used his knowledge of meteorology to inform the public of weather conditions.

Science Background: Weather is the state of the atmosphere at a specific time and place. Changes in weather occur because the *troposphere*, the 10 kilometers of the atmosphere closest to Earth's surface, is constantly moving. The movement is caused by differences in air temperature and air pressure.

Extreme movements in the troposphere cause severe weather. The most common types of severe weather are thunderstorms, tornadoes, and hurricanes.

Thunderstorms are fueled by condensation. When the air at ground level is considerably warmer than the air at a higher level, the movement in the troposphere is dramatic. This movement causes a rapid redistribution of heat. The consequences are rain, thunder, lightning, strong winds, and sometimes hail.

Tornadoes are intense storms with very high winds that circle a small center of low pressure. Tornadoes often form within severe thunderstorms along squall lines.

Hurricanes circle a center of low pressure. They gather energy from the warm ocean waters over which they travel. The warmer the water, the more energy the storm gathers.

I. INTRODUCE

About E. Don Sarreals: Sarreals was a dedicated meteorologist who received national recognition for his work. He published *National Weather Service Forecasting Handbook #2* and several articles.

Ask students what habits of mind they think might have contributed to the success of Sarreals.

About Sarreals' Work: Sarreals was instrumental in the development and success of NEXRAD, also known as Next Generation Radar. The NEXRAD project is supported by the National Oceanic and Atmospheric Administration (NOAA) and by the National Weather Service (NWS). The use of NEXRAD has improved forecasting abilities worldwide.

2. DISCUSS

Use the copy master to prompt discussion. Lead students to conclude that knowing about everyday weather helps us prepare for our daily lives, from choosing our outdoor activities to selecting our wardrobe. Knowing about severe weather helps us stay safe.

You may wish to draw an analogy between satellites used for transmission of cable TV and satellites used for weather forecasting.

Answers to Explain It:
1. Answers will vary. Possible answer: Weather forecasters help you plan activities. They warn you when bad weather is near.

2. Radar tells them what is happening to the air in the atmosphere.

3. Satellites send the signal for the radar images to stations on Earth.

3. EXTEND

After students have completed the Explore It, ask them to share what they have learned. Emphasize the difference between a weather *watch* and a weather *warning*. A watch means that conditions are right for a certain type of weather to be present. A *warning* means that a certain type of weather is present in the area. Discuss the types of severe weather that occur in your area.

Career Focus: The career counselor suggested that if you were interested in becoming a meteorologist, you should also take a course in sign language. Do you agree? Explain your answer.

Oswald S. "Ozzie" Williams

AT A GLANCE

born 1921
Washington, D.C.

education
B.S., M.S., New York University
MBA, St. John's University

occupation
Aeronautical Engineer

Many people turn interests or hobbies into careers. Ozzie Williams did just that.

As a teenager, Williams spent a lot of time making model airplanes. One day he talked to a family friend about how engineers design, or make, airplanes. Williams realized that he could have a career making real airplanes!

Williams went to New York University in the 1940s. The dean there tried to discourage him. He doubted that an African-American man could succeed in the engineering program. Williams, however, went on to earn a master's degree in aeronautical engineering.

During World War II, Williams worked for a company that makes airplanes. While there, he helped design a special airplane. It was the P47 Thunderbolt, a plane that helped win the war.

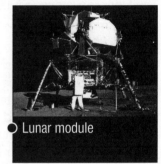
○ Lunar module

During the 1960s, Williams worked on the Apollo space program. Its goal was to send astronauts to the moon. Williams helped build the rocket motors for the Apollo 11 lunar module. In 1969 the engines successfully guided three astronauts to the moon.

? **explain it**

1. What do you think is Ozzie Williams' most important trait? Why?
2. What interests do you have that might turn into a career?

! **explore it**

The Apollo program would not have been successful without the efforts of scientists and engineers like Ozzie Williams. Find out about others who participated in this program. What were their roles and accomplishments?

Oswald S. "Ozzie" Williams

TEACHING NOTES

Purpose: To introduce students to an engineer who helped design the rocket engines used to propel Apollo lunar modules.

Science Background: More than 800 pounds of moon rocks were gathered and brought back to Earth during the lunar missions by the United States and the Soviet Union. Scientists learned about the moon from these rocks. They discovered that the moon is compositionally different from Earth. It has fewer volatile elements—those that boil off at high temperatures. The lunar rocks suggest that, unlike Earth, large parts of the moon were probably molten millions of years ago.

Scientists also learned, from studies of isotopes and elements found in the rocks, that Earth and the moon are approximately the same age—about 4.5 billion years old. The isotopes of oxygen in moon rocks match those of Earth rocks. The implication? The two spheres formed at the same distance from the sun.

These facts and others helped scientists form a new theory about the origin of the moon. In the impact theory, which is now widely accepted throughout the scientific community, a large planetary body collided with Earth about 4.5 billion years ago, ejecting a portion of the debris into Earth's orbit. The debris became the moon.

1. INTRODUCE

About Ozzie Williams: Ozzie Williams is a positive role model for students. Despite discouragement from the dean at New York University, he went on to complete his bachelor's and master's degrees in aeronautical engineering at New York University. He was the second African American to receive a degree in aeronautical engineering.

About Williams' Work: In 1961 Grumman International hired Williams as a propulsion engineer. They were impressed with his knowledge of liquid-fuel rockets. While at Grumman, Williams managed three engineering groups that were responsible for developing the small rocket motors that guided the lunar module of the Apollo 11 spacecraft that landed on the moon in 1969.

Prior to his involvement with Grumman, Williams worked for Greer Hydraulics. There, he helped develop a radar beacon designed to locate downed aircraft.

2. DISCUSS

Use the copy master to prompt discussion. Lead students to understand that a person needs a combination of traits to attain success in any given profession. Discuss some of the traits that might be crucial for a person in engineering, such as tenacity and determination. Explain the importance of such traits when pursuing goals. Also discuss traits for other professions.

Answers to Explain It:

1. Answers will vary. Possible answer: His determination helped him succeed in school despite discouragement from the dean.

2. Answers will vary. Encourage students to be candid with their responses. Help students come up with other career possibilities based on their individual interests.

3. EXTEND

After students have completed the Explore It, ask them to share what they have learned. Students might also like to explore the lives and accomplishments of the astronauts who landed on the moon in 1969 or who were on one of the other Apollo missions.

Career Focus: What kinds of classes do you think would be important for a person who wants to become an engineer?

Dudley Weldon **Woodard**

AT A GLANCE

lived 1881–1965
Galveston, Texas

education
A.B., Wilberforce College
B.S., M.S., University of Chicago
Ph.D., University of Pennsylvania

occupation
Mathematician, Professor

Dudley Woodard grew up during a difficult time for African Americans. Few had good jobs, and schools were legally segregated. This meant that whites and African Americans went to different public schools and received different educations.

Woodard was a determined person, with a supporting family. They encouraged him to get a good education and had faith in him.

Woodard made his family proud. After he attended college in Ohio, he earned a master's degree from the University of Chicago. Woodard then taught mathematics at several schools, including Howard University, which was founded to educate African Americans.

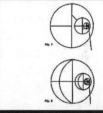

○ **DR. WOODARD's** homeomorphic figures

He enjoyed teaching math. Yet Woodard still wanted to improve himself. So he became a student at the University of Pennsylvania, graduating in 1928. He was the second African American to receive a Ph.D. in mathematics.

Dr. Woodard returned to Howard University. There he developed a graduate program in mathematics and started a mathematics library.

Dr. Woodard retired in 1947. At the time, he was the head of Howard's Math Department.

? explain it

1. What obstacles did Woodard have to overcome to become successful?
2. In what ways could Dr. Woodard be considered a pioneer?

! explore it

Dudley Woodard taught mathematics at Tuskegee Institute, now called Tuskegee University, for several years. Find out about this school. What other important African Americans were involved with this school?

Dudley Weldon
Woodard
TEACHING NOTES

Purpose: To introduce students to one of the first African Americans to earn a Ph.D. in mathematics.

Mathematics Background: Dudley Woodard published the paper "On Two-Dimensional Analysis Situs with Special Reference to the Jordan Curve Theorem," which appeared in *Fundamenta Mathematicae*, 13, in 1929. Formed by French mathematician Marie Ennemond Camille Jordan, the Jordan Curve Theorem states that every simple closed curve, such as a circle, divides a plane into two distinct components, an inside and an outside. Dr. Woodard refined the theorem and proved that it applied to complex closed figures. The Jordan Curve Theorem is relevant to topology, the study of homeomorphisms. Homeomorphisms are figures that can be manipulated to form each other without changing continuity. For example, p and q are homeomorphic, since one can be twisted to resemble the other, but the letters l and 0 are not, since 0 cannot become the letter l without tearing.

1. INTRODUCE

About Dudley Woodard: Dr. Woodard completed his Ph.D. in mathematics at a time during which the separate but equal doctrine of *Plessy v. Ferguson* greatly limited the opportunities available to African Americans. Discuss this historic court decision with students, and help them understand Woodard's incredible accomplishments in light of it.

About Woodard's Work: Woodard spent much of his life teaching mathematics to students at Howard University. His accomplishments there were monumental. He established the master's degree program in mathematics and served as the program's thesis supervisor. In addition, he helped establish a mathematics library. Several of Woodard's students went on to earn their own Ph.D.'s in mathematics.

2. DISCUSS

Use the copy master to prompt discussion. Make sure students understand what a pioneer is. Then discuss Woodard in relation to pioneers in other fields. Ask: What are some of the traits that all of these people probably shared? What conflicts and challenges must pioneers face in their pursuit of their often-lofty goals?

Answers to Explain It:

1. Answers will vary. Possible answer: Woodard overcame the obstacles of segregation and unequal educational and job opportunities.

2. Answers will vary. Possible answer: Woodard could be considered a pioneer because he was one of the first African Americans to receive a Ph.D. in mathematics. He was also the first African American to have one of his papers published in an important mathematics journal.

3. EXTEND

After students have completed the Explore It, ask them to share what they have learned. Students will learn that Booker T. Washington founded the Tuskegee Institute in 1881 to train African Americans in numerous trades. One of the Institute's most famous teachers was George Washington Carver. His work, which led him to develop numerous uses for southern agricultural products, helped bring fame and prestige to Tuskegee.

Career Focus: There are many jobs today that require a good knowledge of mathematics. What are some of these jobs? Which ones interest you the most?

Granville Woods

AT A GLANCE

lived 1856-1910
Columbus, Ohio

education
Informal

occupation
Inventor, Engineer

Granville Woods had little formal schooling. He stopped going to school at age 10. Yet Woods became a successful inventor. How did he do it?

Woods needed to work to help support his family. He became an apprentice in a machine shop and learned the trades of machinist and blacksmith.

In 1872 Woods moved to Missouri, where he worked for a railroad company. While with the railroad, he developed an interest in electricity. He read all he could on the subject. He also took college courses in electrical engineering.

Woods settled in Cincinnati, Ohio, in 1880 and started his own business. His company made many of the products Woods invented.

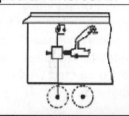

WOODS' railway telegraph system

One of his most important inventions was the railway telegraph system. This system allowed each train engineer to tell whether a train was in front or behind him. Train travel was made safer as a result. Woods also invented a telephone transmitter. It worked so well that the Bell Telephone Company bought it. During his career, Woods invented more than 35 useful products.

? explain it

1. How did Granville Woods learn about electricity?
2. How did Woods make train travel safer?

! explore it

Find out about the history of electricity in this country. When was it first used? What inventors became famous as a result of their work with electricity?

Granville **Woods**
TEACHING NOTES

Purpose: To introduce students to an inventor and electrical engineer who invented more than 35 useful products.

Science Background: Several of Woods' inventions related to the telegraph. Although no one person can be given credit for its invention, Charles Wheatstone, an English scientist, developed the first commercial telegraph line between London and Camden Town, a 1.5-mile stretch, in 1837. Wheatstone's telegraph relied heavily on great advancements in the field of electromagnetism made by Alessandro Volta and Joseph Henry in the early 1830s. In 1844, American inventor Samuel Morse, using many of Wheatstone's ideas, designed a telegraph system that connected Washington, D.C., and Baltimore, Maryland.

The telegraph comprised a battery, a switch, and a small electromagnet. Morse, who is sometimes improperly credited with the invention of the telegraph, was nonetheless important to the device's success. The commercialization of the device by Morse helped spread the technology incredibly fast. By 1866 a line connected the United States and Europe. Morse's greatest contribution was the Morse Code, the language used to send telegraph messages

I. INTRODUCE

About Granville Woods: Woods had to go to work at the age of 10 and never had a chance to finish his grade-school education. Ask students to speculate about how he was able to complete college engineering courses without more formal schooling.

About Woods' Work: Granville Woods' ideas and inventions helped change and modernize transportation and communications. His inventions include the overhead conducting system for electric railways, the third rail for subways, the automatic air brake, the electric egg incubator, the electromechanical brake, a method of tunnel construction for electric railways, the synchronous multiplex railway telegraph, the induction telegraph system, and a galvanic battery.

2. DISCUSS

Use the copy master to prompt discussion. Lead students to the following conclusions: (1) Woods learned about electricity through his work experiences and also through extensive reading. (2) With the induction telegraph system, train engineers could send and receive messages during their trips. If there was a problem anywhere on the route, they could know about it and be able to keep their trains and passengers safe.

Answers to Explain It:

1. Woods developed an interest while working as an engineer for a railroad company. He also read books and took a few college classes in electrical engineering.

2. Woods invented a device that let a train engineer know when a train was ahead or behind him.

3. EXTEND

After students have completed the Explore It, ask them to share what they have learned. You may want to have students make a timeline that shows 10 or 15 events in the history of electricity.

Career Focus: Think about some safety equipment you might invent. How could learning about Granville Woods inspire you to complete your project?

Archie Alexander

As a student at the University of Iowa, Archie Alexander's nickname was "Alexander the Great." Why? It was because of his success as a football player. He loved football, but he also loved his classes in design.

Alexander decided to become an engineer. His adviser at the university had warned him that racism might keep him from getting a job as an engineer. But Alexander stayed with it. At first, he took a small job as a laborer in an engineering company. When his employers realized how capable he was, they made him a bridge engineeer.

○ Pilots at Tuskegee Airfield

Eventually, Alexander and a partner formed their own company. Their company built the heating system and the power plant for the University of Iowa. Soon they became well-known for their work. The partners were hired to design and build bridges, apartments, freeways, and tunnels.

One of Alexander's most notable jobs was building an airfield. During World War II, African Americans came to Tuskegee Institute to become pilots, airplane mechanics, and instructors. Alexander supervised the building of the hangars, control tower, and other facilities. The airfield was a great success.

AT A GLANCE

lived 1888–1958
Ottumwa, Iowa

education
Highland Park College
Cummins Art College
B. S., The University of Iowa

occupation
Engineer, Businessman

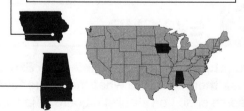

? explain it

1. What did Archie Alexander do to show that he could not be stopped by racism?

2. For what engineering projects was Alexander best known?

! explore it

On a map of Alabama, find Tuskegee. Then, on the Internet or in reference books, find pictures of individual Tuskegee Airmen and, if possible, of the Tuskegee Airmen's planes. Copy the pictures for a bulletin board display.

Archie Alexander
TEACHING NOTES

Purpose: To introduce students to a designer and engineer who was responsible for a variety of structures: bridges, tunnels, freeways, railroad trestles, a power plant, a sewage processing plant, and an airfield.

Science Background: Tuskegee Institute in Alabama, founded in 1881 with a focus on practical learning, introduced courses on aviation as early as 1936. Tuskegee's location was ideal for aviation because the warm climate made year-round flying possible. Also, the rural setting offered undeveloped land for an airfield.

The United States military had been reluctant to admit African Americans into pilot training programs even into the early days of World War II. But pressure from Congress and others finally led to the development of a training program for African Americans at Tuskegee's Moton Field. Engineer/designer Archie Alexander oversaw the construction of the flight school there. The facilities included two aircraft hangars to hold the planes when they weren't in use, buildings for maintenance and storage, and a control tower to oversee the arrival and departure of the planes. The flight training of the 99th Squadron began at Tuskegee in 1941.

By the end of World War II, almost 1,000 African Americans had trained at the Tuskegee Airfield. They flew more than 15,000 missions during the war.

1. INTRODUCE

About Archie Alexander: Before entering the University of Iowa, Archie Alexander attended Highland Park College and Cummins Art College in Des Moines, Iowa. Even with his civil engineering degree, he found that engineering firms were reluctant to hire an African American, so he worked as a laborer for a bridge-building firm. But soon he worked his way into a very responsible position within that company.

About Alexander's Work: In 1914, he and a partner, George Higbee, established their own engineering firm. It soon became recognized for building a heating plant and a power plant at the University of Iowa as well as many other projects—bridges, apartments, viaducts, and tunnels.

In 1925 a tragic accident took the life of his partner. Alexander took on a new partner and the company remained successful. In 1954 President Eisenhower appointed Alexander territorial governor of the U.S. Virgin Islands.

2. DISCUSS

Use the copy master to prompt discussion. Encourage students to check local newspapers for plans of engineering projects, such as new civic buildings, roadways, or parks, to be built near the school or neighborhood. Lead the class in discussing how the plans will change the landscape and will benefit the community.

Answers to Explain It:

1. Alexander took a job as a laborer and worked his way into a professional job.

2. Answers will vary. Possible answer: Alexander was best known for the heating system and power plant at the University of Iowa and for Tuskegee Air Field.

3. EXTEND

After students have completed the Explore It, ask them to share what they have found. Reserve space on a bulletin board or in a display case for students to display the pictures they find of the Tuskegee Airmen and their planes.

Career Focus: The field of engineering has many different specialties. If you were interested in Earth Science, what engineering career would you want to follow? What engineering careers are available to students interested in space?

Guion S. Bluford, Jr.

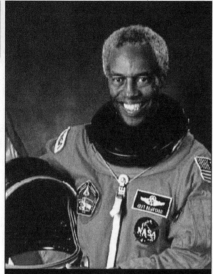

AT A GLANCE

born 1942
Philadelphia, Pennsylvania

education
B.S., Pennsylvania State University
M.S., Ph.D., Air Force Institute of Technology
M.A., University of Houston

occupation
Engineer, Astronaut

Colonel Guion Bluford loves challenges. He was the first African American to travel into space. But Colonel Bluford's career was notable even before he became an astronaut.

At Pennsylvania State University, Bluford majored in aerospace engineering. Those classes helped him understand the mechanics of airplane flight. Being an engineer was just the beginning, though. He also entered flight school and "earned his wings" in 1966. At the university, he joined the Air Force Reserve Officers' Training Corps.

Colonel Bluford flew many combat missions in Asia. He spent five years as a flight instructor. Then he returned to school.

This time, he attended the Air Force Institute of Technology where he studied aerospace engineering and laser science. Toward the end of his studies, he was chosen to become an astronaut. His first mission, aboard the space shuttle *Challenger*, made him the first African American in space. Traveling into space allowed Bluford to bring together all his years of training.

Astronauts in training

In 1987 Bluford earned another master's degree. This time it was in business. Now he's using his experience and training in the aerospace industry.

? explain it

1. In 1966 Bluford "earned his wings." What do you think that means?
2. Why do you think Bluford has a good background to succeed in the aerospace business?

! explore it

On Earth, our bodies are used to the pull of gravity. In space, weightlessness can affect a person's muscles, bones, and blood. Use the Internet to find out how astronauts prepare for weightlessness on space missions.

Guion S. Bluford, Jr.
TEACHING NOTES

Purpose: To introduce students to a highly distinguished astronaut, the first African American to fly into space.

Science Background: Space shuttle *Challenger* was the second shuttle orbiter vehicle to be launched by the United States, but it was the source of many important "firsts" in space. The first spacewalk took place just outside *Challenger* in April 1983. In June of that same year, Sally Ride became the first American woman in space. About two months later, *Challenger* crew member Guion Bluford became the first African American in space. That flight was also the first mission to be launched at night and, later, to land at night.

Challenger completed nine flights, spending a total of 62.41 days in space. It flew 25,803,940 miles altogether, including its final mission. During each of the *Challenger* missions, the crew conducted carefully planned tasks, such as performing scientific experiments and deploying communications satellites. On Guion Bluford's first *Challenger* mission, the crew took medical measurements to understand the physical effects of weightlessness.

On January 28, 1986, *Challenger* exploded 73 seconds after launch. All seven crew members died in that tragedy.

1. INTRODUCE

About Guion Bluford: The list of honors awarded to Colonel Guion "Guy" Bluford gives just a hint of his accomplishments. A Philadelphia native, Colonel Bluford earned a B.S. degree in Aerospace Engineering from Pennsylvania State University in 1964. He went on to earn three graduate degrees, and to log more than 5,200 hours of jet flight time and more than 688 hours in space.

About Bluford's Work: In 1993, after four space missions, Colonel Bluford retired from NASA and the Air Force. He has remained active, however, as a businessman in the aerospace industry and also as a mentor to young people who are interested in pursuing a career in aerospace.

2. DISCUSS

Use the copy master to prompt discussion. Invite students to consider what events in Colonel Bluford's career demonstrate his adventurous, pioneering spirit. Lead students to see that flying combat missions for his country and serving on space shuttle missions required bravery. In addition, he was a pioneer as an African American in the space program.

Answers to Explain It:

1. Answers will vary. Possible answer: It means that Guion Bluford had completed all pilot training and was a full-fledged pilot.

2. Answers will vary. Possible answer: The time he spent in jet flight, in combat, and in space, as well as his master's degree in business, make him uniquely qualified to advise business people in the aerospace industry.

3. EXTEND

After students have completed the Explore It, ask them to share what they have learned. Reserve a portion of a bulletin board or display table for newspaper articles or Internet printouts that provide information on astronaut training and current NASA projects.

Career Focus: Discuss with students what qualities might be important for an astronaut—for instance, resourcefulness, curiosity, and excellent physical fitness. Why would each of these qualities be important to the success of a space mission?

Benjamin Carson

AT A GLANCE

born 1951
Detroit, Michigan

education
B.A., Yale University
M.D., University of Michigan

occupation
Neurosurgeon

If someone had told the young Ben Carson that someday he would be performing brain surgery, he would have laughed at the suggestion. When he was a young student in Detroit, Michigan, Carson did not have much confidence in himself. In fact, he called himself "dummy." His mother, however, believed that he just needed to spend more time studying. As Carson grew older, he realized that she was right. His grades began to reflect his hard work, and he grew up to become a skillful doctor.

At age 33, Dr. Ben Carson was on his way to becoming a world-renowned surgeon. He became the Chief of Pediatric Neurosurgery at Johns Hopkins Children's Center in Baltimore, Maryland. People all over

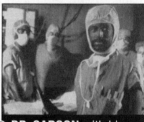

DR. CARSON with his pediatric surgical team

the world recognize Dr. Carson's "gifted hands" and special knowledge. When twin boys with conjoined heads were born in South Africa, the doctors asked Dr. Carson for his help. Dr. Carson went there in 1997 to perform the difficult surgery. The 28-hour operation was a success, even without high-tech equipment. He separated the twins and saved their lives.

Dr. Carson continues to make a difference in the lives of many young people who benefit from his skill and knowledge.

? explain it

1. What does the word pediatric mean?
2. Why do you think Dr. Carson's work requires courage?
3. Who helped Carson gain confidence and become a good student?

! explore it

Computers make it possible for doctors to perform "virtual surgery," rehearsing before they actually operate. Explore some of the other amazing things that computers can do to assist in surgical procedures.

Benjamin Carson
TEACHING NOTES

Purpose: To introduce students to a neurosurgeon who is world-renowned for his success in performing delicate, life-saving surgery on infants and children.

Science Background: Conjoined twins form when a single fertilized egg fails to separate completely into two individuals. In one out of about 2 million live births, twins are born joined at the head. Because each case is different, and because separation is likely to involve extremely sensitive procedures to separate parts of the twins' brains and nervous systems, surgery can be extraordinarily complicated and risky.

With modern medicine and advanced technology, however, some highly skilled surgeons have performed a number of successful operations to separate twins who have been born joined at the head. The development of CT and MRI scanning has made it possible for physicians to see into the living body. These two scanning technologies have been combined in a device called the "Brain Bench," which allows surgeons to gain a 3-dimensional look into the patient's skull to see the configuration of the brain and blood vessels. Using special glasses and "virtual" tools, a surgeon can dissect a virtual brain to see how conjoined twins' blood vessels and nervous systems are intertwined. In this way, adjustments to the procedure can be made before the actual surgery takes place.

1. INTRODUCE

About Ben Carson: When 4-year-old Christopher Pylant of Georgia was diagnosed with a brain stem tumor, his parents were told that he had only 18 months to live. Doctors had indicated that the tumor was inoperable, but the parents consulted a skillful young surgeon, Dr. Benjamin Carson, in the hope that he might be willing to undertake the operation.

The surgery was successful, and Christopher's recovery has become just one of many successes in Dr. Carson's career.

As a youngster growing up in Detroit, Michigan, Ben Carson had little interest in school and very little confidence in his ability to learn. With encouragement from his mother, however, he finished high school and moved on to Yale University, earning a bachelor's degree in 1973. From there, he received a medical degree at the University of Michigan.

About Carson's Work: Dr. Carson's reputation as a neurosurgeon grew as he undertook a delicate operation to separate the German Binder twins. After many months of study and rehearsal, Carson and his team completed the 22-hour surgery with both children surviving.

Dr. Carson and his wife, Candy, work together to manage a foundation that helps hard-working students get funding for higher education.

2. DISCUSS

Use the copy master to prompt discussion. Lead students to consider some of the difficulties that conjoined twins would face, particularly if their heads were joined.

Answers to Explain It:
1. Answers will vary. Possible answer: Pediatrics is the science dealing with the health and growth of children.

2. Answers will vary. Possible answer: The surgery affects the brain. It requires courage to make difficult decisions about these surgical procedures.

3. Answers will vary. Possible answer: Carson's mother supported him.

3. EXTEND

After students have completed the Explore It, ask them to share what they have learned. You might wish to invite a doctor to class to show sample X-ray, CT, or MRI images.

Career Focus: Technology is revolutionizing the field of medicine. Would you like to work in the field of medical technology? Why or why not? What skills would you need?

Name	Date

Dale Brown Emeagwali

AT A GLANCE

born 1954
Baltimore, Maryland

education
B.A., Choppin State University
M.D., Georgetown University
School of Medicine

occupation
Microbiologist, Professor

Dr. Dale Brown Emeagwali loves doing research. That's one reason she became a microbiologist. *Microbiologists* are scientists who study tiny one-celled organisms called microbes. Bacteria and viruses, for example, are microbes. *Microbes* are so small that you need to use a microscope to see and study them. While some of these tiny living things cause illness, others help us.

Dr. Emeagwali carries out laboratory experiments to add to our understanding of microbes. Her special interest is medical research. She is best known for her work with cancer research. On a typical day in the laboratory she first checks up on experiments that are already in progress. Any

EMEAGWALI is a well known cancer researcher.

changes or unexpected developments must be recorded. Next, she sets up equipment and supplies needed for new experiments. Third, she carefully studies all the information that has come from past experiments. Then she writes a report on what she has learned. In this way, she can share the results with others.

In 1996 Dr. Emeagwali was honored as "Scientist of the Year" by the National Technical Association. Yet she feels that the research itself is her greatest reward.

? explain it

1. Why would a microbiologist need a lively curiosity?
2. What is Dr. Emeagwali's workday like?
3. Why do research scientists need to be able to write well?

! explore it

When was the microsope invented? What did early microscopes look like, and how were they used? Investigate how microscopes have changed over the years and how they are used today.

Dale Brown
Emeagwali
TEACHING NOTES

Purpose: To introduce students to a microbiologist who has added to our understanding of cell biology, particularly in the field of medicine.

Science Background: The air we breathe, the food and drink we ingest, our general health, our work and recreation—nearly every aspect of our lives is influenced, in some way, by microbes. While it is true that microbes are responsible for a multitude of diseases, they are also responsible for a great many of life's necessities. In fact, without microbes, the world as we know it would not exist.

Microbes are tiny single-celled organisms. They are so tiny that we cannot see them with the unaided eye. Scientists classify microbes into five groups. *Bacteria*, which we often simply call "germs," cause illness but also break down Earth's wastes and maintain our atmosphere. *Fungi*, including yeasts, help decompose waste but also are vital in producing many of our food products such as bread. *Protists* include amoebas and primitive algae. We recognize *viruses* as the cause of many diseases, but they also benefit us when they are used for laboratory research. *Archaea* are actually living fossils; they help us learn about the earliest forms of life on Earth.

I. INTRODUCE

About Dale Brown Emeagwali: As a child, Dale Brown Emeagwali dreamed of becoming a scientist. Her ambitions kept her at the top of her class. When she is not deeply involved in her work or with her family, Dr. Emeagwali is an accomplished poet; one of her poems has been published in the *Atlantic Monthly*.

About Emeagwali's Work: Being named "Scientist of the Year" by the National Technical Association is just one of Dr. Emeagwali's many accomplishments. As an assistant professor of biology at Morgan State University in Baltimore, Maryland, Dr. Emeagwali shares her interests, skills, and experience with students and with colleagues. In addition, she extends her enthusiasm for science by conducting workshops for inner-city youth.

2. DISCUSS

Use the copy master to prompt discussion. Invite students to an imaginary meal, pointing out how microbes affect the preparation and serving of the food; pizza provides a good example. Begin by explaining why washing hands (to eliminate harmful microbes) is important before preparing or eating food. Then point out how microbes contribute to different parts of the pizza: Yeast makes the dough in the crust rise, microbes are important in the production of the cheese, and soil microbes play a part in the growth of the veggies that top the pizza. You might even mention that microbes will help people digest food and break down any garbage that is left after the meal.

Answers to Explain It:

1. Answers will vary. Possible answer: Curiosity leads microbiologists to ask and investigate many questions in order to discover how microbes affect us.

2. Answers will vary. Possible answer: Dr. Emeagwali's day involves checking on experiments that are already in progress and then planning for new research. She also spends time analyzing data from past experiments.

3. Answers will vary. Possible answer: Researchers need to communicate what they discovered. They share information by writing and publishing reports of their work.

3. EXTEND

After students have completed the Explore It, ask them to share what they have learned. Students who have had an opportunity to use a microscope can describe how they prepared the slide and what they learned from the experience.

Career Focus: Invite a representative from the public health service to visit the class to explain his or her occupation. How can you avoid germs? Discuss practical ways to stay healthy.

Meredith Gourdine

©1998 H.MITCHELL

AT A GLANCE

lived 1929–1998
Newark, New Jersey

education
B.S., Cornell University
Ph.D., California Institute of Technology

occupation
Physicist, Engineer, Inventor

When Meredith Gourdine was young, he was not interested in becoming a scientist. He was a gang member. But he witnessed an event that frightened him into changing his life. His best friend became brain damaged as a result of street violence. To get out of that cycle of violence, Gourdine knew he had to get through high school and go on to college.

During his years at Cornell University, Gourdine was an outstanding track and field athlete, earning him the nickname "Flash." He won a silver medal in the long jump at the 1952 Olympics in Helsinki, Finland.

Computers today rely on Dr. Gourdine's devices to cool microchips.

As an engineering student, he studied the ways in which we use energy. After earning his Ph.D., Dr. Gourdine founded his own company, Energy Innovations. His firm produced devices that convert energy from one form to another without losing heat.

Dr. Gourdine and his company influenced many of the technologies we use today. His ideas helped in the development of photocopiers. Today's computers rely on Dr. Gourdine's project, "heat sinks." These are special devices to cool microchips.

Dr. Gourdine found solutions to many heating and cooling problems.

? explain it

1. Why do you think the nickname "Flash" was a good one for Gourdine?
2. What invention could solve an energy problem in your neighborhood?
3. In what ways could you conserve electrical energy in your school or home?

! explore it

Most of the energy in the United States is generated by burning fossil fuels. Find out where these fossil fuels come from. Make a map showing what you have learned.

Meredith **Gourdine**
TEACHING NOTES

Purpose: To introduce students to a scientist who has made great contributions in engineering physics and energy conversion.

Science Background: Energy consumption in the United States is at crisis proportions. We use much more energy than we can produce. What can we do to solve this problem? What energy sources are available?

Fossil fuels provide much of the energy that we use today. Coal is a major source. Oil, or petroleum, is another fossil-based energy source, though it is not usually used to produce electricity. Instead, energy from oil is generally used by businesses and individuals. Natural gas, considered a fossil fuel by some, is used as a direct source of energy by both homes and businesses. It is a clean form of energy, which means it does not produce much pollution.

What about alternate sources of energy? In many parts of the country, wind energy is being captured by wind-charging devices and then converted into electrical energy. Hydroelectric energy, solar energy, and geothermal energy are also alternate sources of energy.

In order for us to continue using energy at the current level, scientists must come up with new methods of energy production. Meredith Gourdine was one scientist who did just that.

I. INTRODUCE

About Meredith Gourdine: Dr. Gourdine was a dynamic person who overcame many obstacles in life. After beginning his life as a gang member, he found success in athletics and academics. He met each setback with fresh determination and new ideas. Just looking at his set of patents makes it clear that he was a man of unique and innovative ideas.

About Gourdine's Work: Dr. Gourdine was a pioneer researcher in the field of electrogasdynamics, a way of producing high-voltage electricity

from gas. After receiving his Ph.D. from California Institute of Technology, he started his own firm, Energy Innovations.

Dr. Gourdine developed a generator that allowed for cheaper transmission of electricity. He was able to convert natural gas to electricity for every-day use with the principles of electrogasdynamics.

2. DISCUSS

Use the copy master to prompt discussion. Challenge students to list the many ways in which energy is used within the school each day. Prompt them to consider heating and cooling, lighting, computers, televisions, photocopiers and other machines, vehicles used for transport to and from school, maintenance of sporting equipment, and any other relevant energy needs.

Answers to Explain It:

1. An athlete on a track team would need to move as swiftly as a flash.

2. Answers will vary. Possible answer: Encourage creative ideas. Students might suggest, for example, a machine that could capture exhaust gases from cars and trucks and then convert them into usable energy.

3. Answers will vary. Possible answer: Simple ways they can help conserve energy are turning off lights that are not in use, or walking rather than riding to school.

3. EXTEND

After students have completed the Explore It, ask them to share what they have learned. You may wish to divide the class into groups and assign them specific questions to answer, such as: Where is coal mined? Where do we drill for oil in the United States? How much of the oil we use comes from the United States?

Career Focus: Invite students to name several different careers that interest them. What energy uses are related to each of those careers?

Matthew Henson

AT A GLANCE

lived 1866–1955
Charles County, Maryland

education
Honorary degrees from Morgan
State College, Howard University

occupation
Explorer, Navigator

The year was 1909. Explorers Robert Peary and Matthew Henson promised themselves that this year their team would reach the North Pole.

Matthew Henson was the son of free African-American farmers in Maryland. When his parents both died, he left the farm to find work. In Baltimore he got a job as the cabin boy on a ship. The captain taught Henson reading, mathematics, and navigation.

In Washington, D.C., Henson met Robert Peary, who already had traveled into the far north. Peary realized that Henson was skillful as a mapmaker, mechanic, and carpenter. He hired Henson to assist him on many adventurous trips.

HENSON explored the North Pole by dogsled.

In 1899 they set out into the Arctic. They did not reach the North Pole on that first trip. But Henson learned the language and skills of the native Inuits (Eskimos). The Inuits taught him how to break a trail over dangerous ice packs and how to manage a team of sled dogs. He also learned to build an igloo and survive bitter cold.

On April 6, 1909, the two explorers and their team finally reached the North Pole. You can read about that thrilling trip in *Matthew Henson* by Michael Gilman.

? explain it

1. How did Henson's ability to speak the Inuits' language help the team?
2. Why would a mapmaker be especially helpful on a trip into the Arctic?

! explore it

On a world map or globe, find Ellesmere Island, the point from which Peary and Henson set off for the North Pole. Now find the North Pole. Explore why traveling by dogsled between those points could be dangerous.

Matthew A. Henson
TEACHING NOTES

Purpose: To introduce students to an explorer whose knowledge of geographic conditions and navigational skills were critical to the success of the first American team to reach the North Pole.

Science Background: When Peary and Henson set their sights on reaching the North Pole, they were referring to the Geographic North Pole, the northernmost point on Earth. The Magnetic North Pole is one end of Earth's magnetic field. It is the place to which compasses point. It is located about 1,000 miles south of the Geographic North Pole, in Canada. The Geographic North Pole, also called True North, is the point at which all lines of longitude come together, and it is the point at which Earth's lines of latitude begin. The Geographic North Pole is located in the Arctic Ocean. Peary and Henson's team took a sounding at the North Pole and determined that the ocean's depth at that point is more than 9,000 feet. While the pole is usually covered with ice, in recent years, pilots flying over the pole have reported seeing some water.

As might be expected, the climate at the Geographic North Pole is bitterly cold. Surprisingly, though, the North Pole is not the coldest place in the Arctic. That is because the surrounding ocean moderates the climate there.

1. INTRODUCE

About Matthew Alexander Henson: Matthew Henson was a man of versatile skills and interests. As a teenager, he worked aboard a sailing vessel and learned navigation as well as carpentry and basic survival skills. As a young adult, he added salesmanship to his work experience. Henson traveled to Nicaragua with Robert Peary, to investigate the possibility of building a canal between the Atlantic and Pacific.

A few years later he accompanied Peary and four others to the North Pole. Along the way, he learned to build an igloo, manage a dog sled team, speak fluently with local Inuits, and prepare maps of the far north. Henson's skills and experience were critical in bringing the team successfully to the North Pole.

About Henson's work: Henson wrote an autobiography of his travels.

2. DISCUSS

Use the copy master to prompt discussion. Invite students to speculate about the dangers and hardships that a cabin boy might have endured on a sailing ship in the 1860s. Lead students to consider what types of clothing, food, and medical supplies would be available to sailors at that time. Then lead students to discuss how experience on a ship could help a person prepare for a trip to the Arctic.

Answers to Explain It:

1. Answers will vary. Possible answer: By speaking with the Inuits, Henson learned crucial survival techniques.

2. Answers will vary. Possible answer: Henson's mapmaking ability could make it possible for the team to find the camps and food caches they had left for the return trip.

3. EXTEND

After students have completed the Explore It, ask them to share what they have learned. Have a volunteer point out the area in which the explorers traveled to the North Pole. Discuss how cracks in the ice and bitter storms could make travel especially dangerous.

Career Focus: On the chalkboard, write the following job titles: carpenter, mechanic, dogsled manager, Arctic trail breaker, mapmaker, salesperson, igloo builder. Which of those jobs might still be held today? Which jobs do you think would be most interesting?

Fern **Hunt**

AT A GLANCE

● **born** 1948
New York City, New York

education
A.B., Bryn Mawr
M.S., Ph.D., Courant Institute
of Mathematical Sciences

● **occupation**
Research mathematician

Dr. Fern Hunt became interested in mathematics when her ninth grade science teacher encouraged her to participate in science fairs. She is now a mathematician who works at the National Institute of Standards and Technology (NIST). The purpose of the NIST is to help U.S. manufacturers improve the quality of their products.

Dr. Hunt investigates many of the materials that go into U.S. products. She does some of her research by creating models on a computer. For example, to test the glossiness of house paint, she sets up a model of a painted surface on her computer. The model shows how the paint scatters light. The computer "sees" the model just as a person

● **HUNT** gives computer math support.

sees the real thing. From that information, manufacturers can tell how glossy the paint would look and if changes are needed.

Another of Dr. Hunt's projects involves using a computer to find techniques that will make special effects in movies look more realistic.

Dr. Hunt has received awards and honors for her creative use of mathematical techniques. She also enjoys working with young people as a mentor.

(?) explain it

1. How does Dr. Hunt's work show that math allows her to be creative?
2. How might the computer model of a product save time and money?
3. How does Dr. Hunt share her enthusiasm for math research?

(!) explore it

Work with another student to find information about computer animation in movies. Combine the information into one definition for the term *computer animation.*

Fern Hunt

TEACHING NOTES

Purpose: To introduce students to a mathematician whose creative and innovative work demonstrates how mathematical techniques can be applied to improve many aspects of our lives.

Science Background: The National Institute of Standards and Technology (NIST) was established by Congress to assist U.S. industry. Researchers at the NIST give assistance to U.S. manufacturers to help them maintain a competitive edge in the world market. It provides support in many ways. For example, the Institute's researchers conduct investigations on available materials; it supplies databases and publications to provide basic information to manufacturers; it arranges a network of help to small manufacturers; it sets up outreach programs to help with matters such as computer security; and it investigates standards of quality for manufactured goods. The Institute works with the following industry segments: aerospace, automotive, chemical processing, construction, electronic semiconductors, energy conservation, health care, and homeland security. Dr. Hunt's focus at the NIST is to apply mathematical techniques to analyze chemical, electrical, and physical properties of various materials that are used in manufacturing processes. Her work supports the various research projects carried on at the NIST.

I. INTRODUCE

About Fern Hunt: When she was in the ninth grade, a science club adviser showed Fern Hunt how exciting the world of science and math can be. From the Bronx High Scool of Science, she went on to earn an A.B. degree in science at Bryn Mawr and then an M.S. and a Ph.D. from the Courant Institute of Mathematical Sciences.

About Hunt's Work: Dr. Hunt has served as a math professor, as a researcher with the National Institutes of Health, and as a consultant to the Graduate Record Exam mathematics division and the National Bureau of Statistics. In her work as

a math researcher for the NIST, Dr. Hunt provides both analytical and computer math support to researchers who provide important product information to U.S. manufacturers. Dr. Hunt has received many awards and honors, including the Arthur S. Flemming Award for outstanding federal service.

2. DISCUSS

Use the copy master to prompt discussion. Have students pose some questions they would ask Dr. Hunt if they could speak directly to her. Sample questions include: What is it like to work with computer graphics? What are the benefits or difficulties of being a research mathematician?

Answers to Explain It:
1. Answers will vary. Possible answer: Creating a computer model that "sees" as the human eye sees would take a great deal of planning and creativity.

2. Answers will vary. Possible answer: A computer model allows researchers to test materials in the laboratory. These tests can be performed quickly, safely, and inexpensively.

3. Answers will vary. Possbile answer: By serving as a mentor, Dr. Hunt shares her enthusiasm for math with young people.

3. EXTEND

After students have completed the Explore It, ask them to share what they have learned. Let students work together in small groups to find information about computer animation. If possible, let each member of the group use a different source; then have the group blend the information into one definition.

Career Focus: Name five careers that seem interesting to you. How is a having a knowledge of math important in order to succeed in each career?

Trachette L. Jackson

When Trachette Jackson entered college, she planned to become a doctor. She had a passion for biology. Then, in her senior year, a contest changed her career plans. The math department at her university held a contest involving a problem from biology. Dr. Jackson entered and won the contest. In the process, she realized how important math can be in solving medical problems.

Dr. Jackson's research focuses on cancer cells. With her strong background in science and math, she is able to develop computer models.

Computer modeling helps cancer research.

How can a computer model answer questions about cancer? Using mathematical equations, Dr. Jackson is able to represent cancer cells on the computer. Applying mathematical techniques, she can show how the cells grow and develop. The model can predict how cancer cells will respond to different conditions. By experimenting with the model, Dr. Jackson is able to suggest ways in which cancer could be treated successfully. Her research adds valuable information to medicine's quest for a cancer cure.

Today, she is a teacher, lecturer, and author.

AT A GLANCE

born 1972

education
B.S., Arizona State University
M.S., Ph.D., U. of Washington

occupation
Mathematician

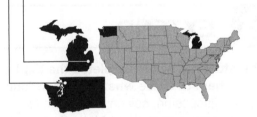

? explain it

1. How do you think Dr. Jackson's studies in biology help in her research?
2. How can computer modeling help doctors treat people's illnesses?
3. Why is it important for a researcher to be able to give speeches?

! explore it

Dr. Jackson does research to learn about cancer tumors. In the nineteenth century, Marie Curie also did research to learn about tumors. Find out about Marie Curie's research.

Trachette L. Jackson
TEACHING NOTES

Purpose: To introduce students to a scientist and mathematician who uses computer modeling and research to help solve medical mysteries.

Science Background: Computer modeling is a research tool that helps scientists make educated predictions. In industry, it is being used to predict how various materials can endure time and the environment. In meteorology, it is used to predict the weather. In anthropology, it is sometimes used to recreate three-dimensional images of ancient peoples and artifacts. Information obtained with computer modeling is often combined with other research data to gain a full understanding of the problem under investigation.

In medicine, computer modeling has many different applications. It has been used to "practice" delicate surgeries before they are actually performed. It has been used to predict the effects of certain treatments on diseases. It is even used to help strengthen our community by modeling how an infectious disease outbreak might occur. After entering the appropriate mathematical information, researchers can set the computer in motion and observe the likely progression of a disease, along with the effect of community response to the problem.

I. INTRODUCE

About Trachette Jackson: Although she is young, Dr. Trachette Jackson has already achieved a great deal in her career. After earning a B.S. with high honors at Arizona State University, she earned an M.S. and then a Ph.D. in applied mathematics at the University of Washington in 1998.

About Jackson's Work: Dr. Jackson has worked as a visiting scientist at the National Health and Environmental Effects Research Laboratory. Now, as an associate professor of mathematics at the University of Michigan, she does research applying mathematics to biomedical systems, particularly tumor biology.

2. DISCUSS

Use the copy master to prompt discussion. Point out that computer modeling is one technique that scientists and mathematicians can use when direct observation is not practical. For instance, finding out how weather and time will affect the finish on a car could take many years, but a computer model can predict the results in a much shorter time.

Emphasize that mathematics affects our lives in many ways and that it is used for more than just everyday calculations and measurements.

Answers to Explain It:
1. Answers will vary. Possible answer: Dr. Jackson's knowledge of different body systems helps her know how to set up experiments for computer modeling.

2. Answers will vary. Possible answer: Computer models can display information that helps doctors predict how different treatments will affect a disease.

3. Answers will vary. Possible answer: Giving lectures is one way of sharing new information so that others can make practical uses of it.

3. EXTEND

After students have completed the Explore It, ask them to share what they have found. In class discussion, call on volunteers to explain what they have read about Marie Curie and her research. Make sure that everyone understands that Curie's original intent was to find a cure for tumors, but her work with radioactive materials led to other innovations, including X-ray machines. By discussing the conditions under which Marie Curie and Trachette Jackson work, emphasize that available technology influences the type of investigation that can be done.

Career Focus: What are some of the classes that a student must take in order to become a doctor? Which of those classes might involve doing some research in a laboratory?

Raymond L. Johnson

AT A GLANCE

born 1943
Alice, Texas

education
B.S., University of Texas
Ph.D., Rice University

occupation
Mathematics professor

What's behind computers, DVDs, and technological advances in every field? The answer is math. For example, mathematical functions are used to indicate how a computer screen changes colors. Math makes it possible to store an entire movie, along with extra information, on a DVD. Dr. Raymond Johnson's research area deals with interesting matters such as these.

Getting to a career in advanced mathematics was not easy. Racial discrimination made life difficult for African-American students attending universities in the 1960s. Classes were integrated, but dormitories, sports, and campus activities were still segregated. Dr. Johnson concentrated on his studies, though, and became the first African-American student to earn a Ph.D. from Rice University.

Math research helps encode fingerprint files.

Dr. Johnson continues to teach and do research at the University of Maryland. Remembering the difficulties that he faced, he works to eliminate barriers for African-American students.

? explain it

1. What racial barriers did Dr. Johnson overcome as a student?
2. How did his experiences with discrimination shape his career?
3. Why do American students need to learn all they can about mathematics?

! explore it

When you add, subtract, multiply, or divide, you are calculating. The word "calculate" has an interesting history. Use an unabridged dictionary to find out what language it comes from and its meaning in its original language.

Raymond L. **Johnson**
TEACHING NOTES

Purpose: To introduce students to a university professor who, in addition to doing research and teaching mathematics, works actively to recruit and maintain minority students in his university's graduate math programs.

Mathematics Background: A mathematical analyst advances our knowledge by examining mathematical functions and the relationships between them. The analyst breaks a problem into its component parts and then uses a variety of techniques—computer modeling, for instance—to examine each of the components. The new data can then be examined and interpreted.

Harmonic analysis, one of Dr. Johnson's special interests, has interesting and useful applications. For instance, in law enforcement, harmonic analysis gives a very compact way to encode fingerprint files. Harmonic analysis has also made electronic technology more efficient by allowing larger amounts of information to be stored on smaller files.

I. INTRODUCE

About Raymond Johnson: The restless 1960s influenced Dr. Raymond Johnson's education in ways that nobody could have predicted. Although he was not allowed to attend a new all-white neighborhood school, he benefited from the nation's urge to promote science and math training following Russia's launch of Sputnik and was placed in accelerated science and math classes in an African-American school.

During his university years, the spirit of the times influenced Dr. Johnson's education again. Minority students stayed in segregated dormitories and were not allowed to participate in sports. Later, Dr. Johnson became the first African American to study at Rice University.

About Johnson's Work: Dr. Johnson is a professor at the University of Maryland. From 1991 to 1996, he served as chair of the Mathematics Department there. As a researcher, he has a special interest in harmonic analysis. As an educator, he has a special interest in helping minority students succeed in graduate math programs.

2. DISCUSS

Use the copy master to prompt discussion. Lead students to consider a classroom, like the ones in Raymond Johnson's early school, with students at four different grade levels. Could students in first, second, and third grades all share games and sports together? Would they share the same interests for science experiments? What would be some of the advantages? Could the fourth-grade students help first graders learn to solve math problems or practice reading?

Answers to Explain It:
1. Answers will vary. Possible answer: Early in his education, Raymond Johnson was not allowed to attend the new all-white neighborhood school. While in college he faced racial barriers such as segregation in the dormitories and sports activities.

2. Answers will vary. Possible answer: Dr. Johnson works to help minority students succeed in college math programs, particularly at the graduate level.

3. Because we use math in some form nearly every day, and because we rely more and more on technology that depends on math, we need to know as much as we can about it.

3. EXTEND

After students have completed the Explore It, ask them to share what they have learned. Calculate comes from the Latin word for "stone." In the days of the Roman Empire, people used small stones for calculations. Students may also be interested in learning the word histories of "add," "subtract," and "mathematics."

Career Focus: What are some of the many ways in which your work as a mathematician could influence lives?

Percy Julian

AT A GLANCE

lived 1898–1975
Montgomery, Alabama

education
B.S., DePauw University
M.S., Harvard University
Ph.D., University of Vienna, Austria

occupation
Chemist, Inventor

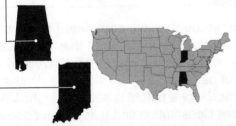

Percy Julian liked challenges. Even as a young person, he was a good problem solver. For instance, he faced a problem early in his education. African-American students in his city were not allowed to attend school after the eighth grade. Julian did not let that stop him. He entered DePauw University knowing that it would be difficult to catch up. He studied hard and graduated at the top of his class. From there, he continued to study and to teach chemistry.

Julian synthesized many useful medicines from soybeans. In 1935 he developed the drug physostigmine, used to treat glaucoma, an eye disease. Later, his research produced a substance that was

●**DR. JULIAN** was responsible for 100 chemical patents.

helpful in treating cancer. In 1949 Dr. Julian's team developed a medicine that would relieve arthritis. The drug that had been used for that purpose was extremely expensive. Dr. Julian's soybean-based substitute cost only pennies per ounce.

Besides solving medical problems, Dr. Julian invented other useful products from soy. He found a way to make printers' inks from soybeans. He developed a foam that put out oil and gas fires that was used in World War II.

? explain it

1. How did Dr. Julian's work help soybean farmers?
2. Explain how Dr. Julian's work saved some lives and improved others.
3. Do you think inks made from soy are Earth-friendly? Why?

! explore it

Use the Internet to find out about patents. Why would Dr. Julian want to obtain patents for his inventions?

Percy Julian
TEACHING NOTES

Purpose: To introduce students to a chemist and inventor who synthesized a number of important medicines as well as Earth-friendly inks and a fire-retardant foam, all from soybeans.

Science Background: Soybeans were grown in China as far back as 5,000 years ago as an important food source. In fact, they were considered to be a "sacred" grain. They were first brought to the United States in 1804. A group of sailors used the soy beans to weigh their boat down and properly balance it.

By 1829 some U.S. farmers were growing soybeans to be used as cattle feed. During the Civil War, when coffee beans were scarce, roasted soybeans were used occasionally as a substitute.

During the 1900s, George Washington Carver and Dr. Percy Julian began exploring the possibilities for synthesizing food and other products from soybeans. Today, soybeans are used as a major source of oils and protein throughout the world. Many food products include soy protein as an ingredient. The United States produces more than 2 billion bushels of soybeans a year, a large portion of which is exported.

1. INTRODUCE

About Percy Julian: The son of a mail clerk, Percy Julian graduated from DePauw University in 1920. Although Dr. Julian was a gifted student, chemist, and inventor, he suffered the effects of lingering racial prejudice. His professors at DePauw feared that he wouldn't be able to get a job that reflected his abilities.

About Julian's Work: Dr. Julian taught chemistry at Howard University and DePauw. He left teaching when he joined the Glidden Company in 1936. He stayed there until 1954, when he left to open his own research labs. Dr. Julian received many awards and honors including 19 honorary doctorate degrees. In 1947 he received the

Spingarn Medal from the National Association for the Advancement of Colored People (NAACP).

2. DISCUSS

Use the copy master to prompt discussion. Help students explore the concept of racial prejudice by asking them to think about what it would be like to be treated badly because of personal characteristics. Discuss the following with students: Dr. Julian had to turn down a position at a college in Wisconsin because the town wouldn't let an African American stay overnight. How do you think this affected him?

Answers to Explain It:
1. Answers will vary. Possible answer: By finding many ways in which soybeans can be used, Dr. Julian's work increased the need for farmers to grow more soybeans.

2. Answers will vary. Possible answer: The medicines that he synthesized, which were more affordable, were used to treat people with serious illnesses.

3. Answers will vary. Possible answer: Products made from petroleum use up our limited resources, while soy-based products are made from a renewable resource.

3. EXTEND

After students have completed the Explore It, ask them to share what they have learned. Have students share what they have learned about patents. Students should understand that the basis for a patent is set forth in the United States Constitution and that the purpose is to encourage artists and inventors.

Career Focus: Suppose you are a registered ophthalmologist. Your office is located in a hospital wing. There is a lot of support to name the wing after the hospital president. You have been asked to give your opinion. Would you feel comfortable suggesting that the new wing be named in honor of Percy Julian? Why? Why not?

Dorothy McClendon

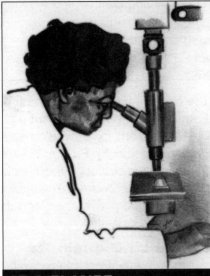

How is an army tank like an egg salad sandwich? The answer may be surprising. Both of them can be completely ruined by microbes. Microbes—bacteria and fungi, for example—are living organisms too tiny to be seen by the unaided eye.

Dorothy McClendon, a microbiologist, investigated microbes in army fuel tanks. She coordinated microbial research for the Research and Development Division of the U.S. Army Tank and Automotive Commission.

McClendon found that microorganisms can break down the fuels that are used in tanks. Spoiled fuel can damage the sensitive moving parts of vehicles, destroying the entire tank. The army needed to find ways to keep fungi and other microbes from breaking down the fuel in tanks.

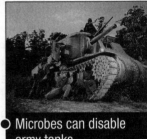

○ Microbes can disable army tanks.

McClendon was well suited for that work. As a former teacher, she had shared her knowledge about microbes and their behavior with her students. Her knowledge, along with her eye for detail and her questioning mind, were valuable qualities for the research. She helped develop methods for preventing microbes from spoiling the fuel in fuel tanks and in military storage depots.

AT A GLANCE

born 1924
Minden, Louisiana

education
B.S., Tennessee A & I State University
Wayne State University, University of Detroit

occupation
Microbiologist

? explain it

1. Why would McClendon need a microscope for her work?
2. How did McClendon's teaching help her do research?
3. How does hand washing prevent illness?

! explore it

When an oil tanker ship leaks oil into a waterway, great damage is done to the water and to the wildlife that live there. Find out how microbes may be used to help clean up oil spills.

Dorothy McClendon
TEACHING NOTES

Purpose: To introduce students to an industrial microbiologist whose skills have been valuable to the U.S. Army.

Science Background: Military combat during the 1900s led to the development of armored vehicles known as tanks. Most of these vehicles carry cannons and machine guns into battle. A gun turret on top of the vehicle can swivel, allowing a wide range of motion. On level ground, tanks can travel at 50 miles an hour. They also are capable of climbing over rough terrain, including hills and slopes.

Machines with these capabilities need highly efficient lubricants and fuel while they are in operation. Even when tanks are in storage, careful attention must be paid to their fluids and lubricants. The growth of fungi or other damaging microbes could break down the fuel, leading to serious vehicle damage.

As the coordinator for microbial research at the U.S. Army Tank Automotive Command (TACOM), Dorothy McClendon focused on developing fungicides that would protect tanks' fuel oil and other stored material from microbial damage.

I. INTRODUCE

About Dorothy McClendon: The town of Minden, in northwest Louisiana, is surrounded by piney hills and beautiful waterways. Growing up in this environment, Dorothy McClendon developed a special interest in wildlife. When she moved to Detroit, Michigan, she carried that interest with her, majoring in biology at Cass Technical High School. She attended Tennessee Agricultural and Industrial State University. Teaching was a specialty of the school, and Dorothy McClendon graduated in 1948 with a major in teaching biology. She taught in Arizona and Arkansas.

About McClendon's Work: A new and rather different chapter in McClendon's life opened when

she became a microbiologist for the U.S. Army Tank Automotive Command in Warren, Michigan. As a professional microbiologist for 24 years, McClendon led the Research and Development team in investigating the effect of microbes on vehicles that were essential to U.S. defense.

Today, McClendon is retired, but she is honored as an outstanding teacher and research scientist.

2. DISCUSS

Use the copy master to prompt discussion. Invite students to speculate about the characteristics that a microbiologist would need to complete work successfully. Prompt them to point out that using a microscope requires patience, that pursuing the answers to research questions takes curiosity and persistence, and that keeping an open mind is essential.

Answers to Explain It:

1. Answers will vary. Possible answer: McClendon studied living things so tiny that most of them can't be seen with the unaided eye.

2. Answers will vary. Possible answer: As she prepared lessons and shared information with students, she increased her own background knowledge about microbes and how to do research.

3. Answers will vary. Possible answer: Disease-causing microbes can gather on our hands through normal daily activities. Washing with soap can reduce bacteria and therefore the risk of infection.

3. EXTEND

After students have completed the Explore It, ask them to share what they have learned. Have students look for news articles about oil spills in U.S. waters. Call on students who have located information about microbial control of oil spills to explain their findings to the class.

Career Focus: How many occupations related to microbiology can you name? What medical or food services careers would require a knowledge of microbiology?

Camellia Moses Okpodu

AT A GLANCE

born 1964
Portsmouth, Virginia

education
Ph.D., North Carolina State
University

occupation
Chairman, Department of Biology,
Norfolk State University

Dr. Camellia Okpodu is very interested in plantlets. These are plants that are developed in a laboratory. For instance, strawberry plants are able to make copies of themselves. The plant sends out long shoots that grow roots into the soil. Then a perfect copy of the parent plant can develop from the new root. The new plant is called a *plantlet*. Dr. Camellia Okpodu and other plant physiologists conduct experiments to discover the best conditions for growing such plantlets.

Why would we want some plants to make perfect copies of themselves? Disease, drought, or insects can destroy a valuable crop. To produce successful crops, farmers need plants strong enough to resist those

○ **OKPODU** developed disease-free oak trees.

dangers. Copying hardy plants with desirable fruits boosts our ability to grow food. Copying shade trees and other plants that add beauty to our environment is important, too. Dr. Okpodu and her students have focused their work on developing disease-resistant oak trees.

As chairman of the Biology Department at Norfolk State University, Dr. Okpodu seeks ways to help plants grow. She is also very interested in helping to increase students' knowledge of plants.

? explain it

1. What do you think *physiologist* means?
2. Why do scientists need to keep a plant laboratory free of germs?
3. Why might some laboratory experiments with trees take many years?

! explore it

Grow your own plant copy. Place a leaf or piece of stem from a house plant, such as a spider plant, into clean water. Give the plant light. When the roots are well developed, transfer to a pot with well-drained soil.

Camellia Moses
Okpodu
TEACHING NOTES

Purpose: To introduce students to the first African-American woman to earn a Ph.D. in plant physiology.

Science Background: A hungry world needs to produce strong, disease-free crops and healthy plants. In a laboratory, scientists can use a controlled environment to produce precise copies of some plants that have been found to have desirable characteristics, such as high yields; drought-, disease-, or insect-resistance; and appealing taste. The process is called *vegetative propagation* or *plant tissue propagation*.

In plant tissue propagation, a small portion of the parent plant—a budded stem, a leaf, a node, or a root segment—is removed and placed, under sterile conditions, into a cultivating medium. Nourished by the growing medium, the tissue develops into a "plantlet" with roots and stem. In contrast to plants grown from seed, these plantlets develop from just one parent plant. Since they have the same genetic makeup as the single parent plant, they grow into mature plants with the same characteristics as the parent. Not all plants can be duplicated in this way, but particularly desirable strains of strawberries, potatoes, and even some trees have been developed successfully through plant tissue propagation.

1. INTRODUCE

About Dr. Camellia Moses Okpodu: Dr. Okpodu's interest in growing plants developed early in life, on her grandparents' farm in North Carolina. After high school, she attended North Carolina State University. It was there she discovered that her real enthusiasm lay in learning about plant biology. Study in her field was rewarding and interesting, but sometimes she found it to be difficult because of financial constraints and racial prejudice. After years of hard work, Dr. Okpodu became the first African-American woman ever

to earn a Ph.D. in plant physiology and biochemistry. Today she is chairman of the Department of Biology at Norfolk State University.

About Okpodu's Work: Dr. Okpodu's work focuses on producing plants that can resist environmental threats such as disease, drought, and insects. Even more important, her work focuses on helping young people share her enthusiasm for, and experience with, plant development.

2. DISCUSS

Use the copy master to prompt discussion. Help students understand that laboratory work requires patience and precision. Challenge students to think of ways in which information from plant experiments can benefit them (e.g., by helping to produce fruits and vegetables with exceptionally good taste and nutritional value).

Answers to Explain It:

1. A *physiologist* is a scientist who studies living organisms and their parts.

2. Answers will vary. Possible answer: A plant disease could change or even destroy the plants being studied in an experiment.

3. Answers will vary. Possible answer: Since some trees take many years to grow to maturity, experiments with trees must be conducted over long periods of time.

3. EXTEND

After students have completed the Explore It, ask them to share what they have learned. You might want to have each student keep a record, over a period of several weeks, of the growth and development of his or her plant, from the early stages to full development.

Career Focus: Students with an agricultural background can share their experiences with classmates. What kind of work goes into raising a successful crop? Do you think that careers with plants can be found in urban, suburban, and rural settings?

Ashanti Johnson Pyrtle

AT A GLANCE

born 1970
Dallas, Texas

education
B.S., Texas A&M University,
Galveston
Ph.D., Texas A&M University

occupation
Oceanographer, Professor

Sharks are exciting, but they might not seem inviting to everyone. Ashanti Johnson Pyrtle, however, became interested in them in third grade. Her school project on sharks and dolphins made her realize that she wanted to become a marine scientist. As a teenager, she volunteered at the Dallas Public Aquarium.

Ashanti Johnson Pyrtle received her Ph.D. in oceanography. She became interested in learning how chemicals affect ocean plants and animals. Her research now focuses on radioactive pollution in water systems. Some *radionuclides*, radioactive substances, occur naturally. Others have entered Earth's water from nuclear reactors and nuclear tests. Dr. Pyrtle and her team study

PYRTLE testing for radioactive pollution

where these pollutants are found, how they are transported, and how they affect our oceans.

The health of the oceans has an effect on the health of land plants and animals. Dr. Pyrtle's research helps us learn how the health of the oceans will affect us, too. She and her team have worked in the Russian Arctic as well as the southern United States.

Dr. Pyrtle enjoys sharing her enthusiasm for oceanography. She is active in helping young people become involved in marine studies.

? explain it

1. What duties might a volunteer at a public aquarium perform?
2. What are some ways in which oceans could become polluted?
3. Why is it important to study radioactive pollution?

! explore it

On a map of the United States, find the Savannah River. Dr. Pyrtle has done some of her research there. If pollution from an old nuclear generator were to seep into the Savannah River, how might it get into the Atlantic Ocean?

Ashanti Johnson Pyrtle
TEACHING NOTES

Purpose: To introduce students to an oceanographer who studies radioactive contamination in water systems and who gives important assistance to young people interested in studying oceanography.

Science Background: A radionuclide, or radioactive nuclide, is an atom with an unstable nucleus. Some radionuclides, such as radon, occur naturally. Others are produced when atoms split apart, as in the atmospheric nuclear weapons testing that occurred during the 1960s.

Today, we benefit from radionuclides in a number of ways. Some of our medical equipment and many of the technologies that produce our food and drinking water rely on radionuclides. If radionuclides escape into the environment through careless disposal, accident, or other ways, however, they can pose a real danger from radioactive contamination.

Radioactive cesium-137 is known to have contaminated the soil in some areas, for instance, where nuclear accidents have occurred. Precipitation and erosion can bring that contamination to nearby bodies of water. Dr. Pyrtle and her group have investigated the movement and behavior of these radionuclides in the Lena River, near the Chernobyl accident, and also in the United States.

I. INTRODUCE

About Ashanti Johnson Pyrtle: In addition to her investigation of the distribution and behavior of radionuclides, Dr. Pyrtle has also devoted herself to helping people of all races and backgrounds realize the goal of studying oceanography. As a student, Ashanti Johnson Pyrtle noticed with surprise that there were few minorities in this field, and she is now determined to support and encourage young people's interest in marine science.

About Pyrtle's Work: Among her many activities to assist students, Dr. Pyrtle is the director of the Minorities Striving and Pursuing Higher Degrees of Success in Earth System Initiative. She also co-directs the USF College of Marine Science OCEANS GK-12 Fellowship Program.

2. DISCUSS

Use the copy master to prompt discussion. Lead students to consider how nuclear contamination as far away as the Arctic Ocean can affect the lives of everyone on Earth. For instance, keeping the oceans free of contamination supports the fishing industry, an important source of food.

Answers to Explain It:

1. A volunteer might assist by helping visitors find their way to particular exhibits. Volunteers might also help feed and care for marine animals.

2. Answers will vary. Possible answer: Oil transporters need to be careful to avoid spills. Coastline cities need to prevent waste from seeping into the oceans.

3. Answers will vary. Possible answer: The study of radioactive pollution helps keep our oceans free of contamination and marine life safe.

3. EXTEND

After students have completed the Explore It, ask them to share what they have learned. Dr. Pyrtle has investigated radioactive contamination in the Lena River, near the site of the Chernobyl nuclear accident. Interested students can locate Chernobyl and the Lena River on a world map to visualize the area in which her research took place.

Career Focus: Many different types of studies are open to oceanographers, including: investigating wave and tide action, studying marine plants and animals, and learning about the geology of the ocean floor. What kinds of courses would you need to prepare for a career in oceanography? Why would the study of foreign languages be recommended?

Charles H. Turner

AT A GLANCE

lived 1867-1923
Cincinnati, Ohio

education
B.S., M.S., University of Cincinnati
Ph.D., University of Chicago

occupation
Zoologist, Educator

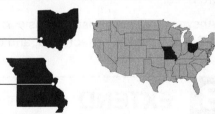

Can bees see color? How do ants tell each other where to find food? Charles Turner was always interested in observing insects. He investigated some of these questions that help us understand insect behavior. Why is this important? Humans share the earth with at least 800,000 species of insects.

After completing his university studies, Dr. Turner decided that he wanted to share his curiosity and knowledge with young people. He traveled to St. Louis, Missouri, to teach at Sumner High School. He took his students on field trips and taught them to observe and record all they saw.

Some of Dr. Turner's most fascinating work concerned bees and ants. He was one of the first

Colorful flowers attract bees.

to realize that insects could see certain colors. Have you ever noticed that bees visit red flowers more than they do any other color? Using that information, beekeepers can plant flowers that tend to attract bees. In addition, Dr. Turner detected a behavior pattern in ants that no one else had ever noted. Before ants return to their nests, they perform a circling motion, called *Turner's circling*. Though we do not yet know the reason for this, being aware of the behavior helps us ask questions for future study.

? explain it

1. Why do you think learning about the way insects behave is important?
2. How could beekeepers use the information that bees prefer red flowers?
3. What are some process skills that scientists like Turner use?

! explore it

Have you ever seen an insect running over the top of a pond or marsh? Use the Internet or a reference book to find out about the water strider.

Charles H. Turner
TEACHING NOTES

Purpose: To introduce students to a scientist whose experiments and observations contributed to our understanding of insect behavior.

Science Background: Insect behavior is controlled by many factors, including instinct. As Charles Turner demonstrated, however, not all insect behavior is instinctual. By teaching bees to fly through a maze made up of different colors and shapes, Dr. Turner showed that bees can learn. Other types of insect behavior are regulated by chemicals automatically produced within the bodies of the insects. These chemicals are known as *pheromones* (scents) and are used to communicate with other bees. For instance, one honeybee's pheromone warns others of danger to the hive. If some type of threat is detected, the honeybee emits a pheromone that alerts others in the same colony of the need to respond. Honeybees also respond to a specialized pheromone produced by the queen bee. Each queen has her own specific pheromone, which the worker bees detect to identify their hive. When she is near, the queen's pheromone causes worker bees to turn and face her, and give her whatever attention is needed.

As entomologists learn more about insect pheromones, we can better control unwanted or harmful insects. Scientists have created artificial pheromones that attract male insects into a trap so farmers can reduce their use of pesticides.

1. INTRODUCE

About Charles Turner: Dr. Turner was a dedicated and inspiring teacher. His interest in biology led him to earn both a bachelor's and a master's degree in biology at the University of Cincinnati. After teaching there for a while, he traveled to the University of Chicago and earned a Ph.D. in biology in 1907. Realizing that he particularly enjoyed working with high school

students, he took a teaching position at Sumner High School in St. Louis, Missouri. There he continued his experiments with animal and insect behavior.

About Turner's Work: Dr. Turner published more than 50 journal articles on animal behavior, neurology, and invertebrate ecology. Some of his more notable papers include "Psychological Notes on the Gallery Spider," "Experiments on the Color Vision of the Honey Bee," and "Do Ants Form Practical Judgments?" In the course of his research, he found that many invertebrates, such as moths and cockroaches, can be trained to respond to specific stimuli, much like the dogs in Pavlov's experiments.

2. DISCUSS

Use the copy master to prompt discussion. Lead students to note that farmers make practical application of entomologists' findings to prevent insect destruction of their crops. Also public health officials use knowledge of some insects to prevent the spread of infection.

Answers to Explain It:

1. Answers will vary. Possible answer: Knowing the behavior of insects helps farmers.

2. Answers will vary. Possible answer: Beekeepers can plant red flowers to help them attract more bees.

3. Answers will vary. Possible answer: Scientists like Turner observe, collect data, and compare.

3. EXTEND

After students have completed the Explore It, ask them to share what they have learned. Encourage them to share Internet sites and print sources that give particularly good information about insect behavior.

Career Focus: Discuss with students the variety of careers available to people interested in insects. Suppose you work for an exterminating company. How would understanding how insects act help you do your job?

Neil deGrasse Tyson

AT A GLANCE

born 1959
Bronx, New York City, New York

education
B.A., Harvard University
M.A., University of Texas at Austin
Ph.D., Columbia University

occupation
Astrophysicist; Director, Hayden
Planetarium

When 9-year-old Neil Tyson first visited the Hayden Planetarium in New York City, he never imagined that someday he would be the director there. He was already very curious about the stars and the sky. It is also true that the planetarium provided the exhibits and information that allowed his imagination to fly. But at that time, Tyson and his friends spent most of their spare time playing sports.

When he got to college, Tyson decided to become an astrophysicist. An astrophysicist is a scientist who applies the laws of physics to the stars and planets.

Dr. Tyson is gifted at explaining complicated concepts and making them fun and easy to understand. How does he do that? First, as

TYSON shows his enthusiasm for science.

the director of the Hayden Planetarium, he oversees the exhibits and programs that bring enthusiasm about the universe to the visitors. Dr. Tyson also travels and gives lectures. He has written many books and has conducted a television series about astronomy.

If anyone advises you to "reach for the stars," think of Dr. Tyson. With hard work and study, he has reached for the real stars and has even become a media star himself.

? explain it

1. Why might it be difficult for a person in a large city to observe stars?
2. What would you expect to find in a planetarium?
3. Why do we use "star" to refer to people in sports and entertainment?

! explore it

How much do you know about dwarf galaxies? Find out more about them, and write five questions about dwarf galaxies that you think a researcher might investigate.

Neil deGrasse Tyson
TEACHING NOTES

ten or co-written a number of books including *Origins*, *The Sky Is Not the Limit*, and *Merlin's Tour of the Universe*. In his writing, Dr. Tyson applies familiar, everyday concepts to help readers understand the complex and sometimes bewildering universe.

Purpose: To introduce students to an astrophysicist who is the youngest person ever to serve as director of the Hayden Planetarium in New York City.

Science Background: Dwarf galaxies, a special interest of Dr. Tyson's, are small clusters of stars just beyond our Milky Way Galaxy. The Sagittarius Dwarf Elliptical Galaxy, also called SagDEG, was discovered in 1994. At the time scientists considered it to be our nearest neighbor. Then, in November 2003, a closer dwarf galaxy, Canis Major Dwarf Galaxy, was discovered through the analysis of data, rather than through direct observation. Infrared images of this dwarf galaxy are now available.

Early in 2004, 12 astronomers working together in Australia announced that they had discovered 40 dwarf galaxies. These clusters are so small that they appeared at first to be single stars. Scientists suggest that these dwarf galaxies may have joined together from bits of material left over when the large galaxies were formed. The newly discovered galaxies are about 60 million light years away from Earth. Clearly, investigation of dwarf galaxies will be an interesting and exciting topic for the future.

1. INTRODUCE

About Neil deGrasse Tyson: Born and raised in New York City, Dr. Tyson received his Ph.D. from Columbia University. Dr. Tyson's energy and enthusiasm have made him a well-known media figure with the ability to explain the workings of the universe in simpler terms for the average person.

About Tyson's Work: Dr. Tyson is astrophysicist and director of the Hayden Plantetarium at the American Museum of Natural History. His work focuses on the characteristics of galaxies. In addition to frequent lectures and his appearances on the television show *NOVA*, Dr. Tyson has writ-

2. DISCUSS

Use the copy master to prompt discussion. Comment that in his autobiography, *The Sky Is Not the Limit*, Dr. Tyson discusses some of the difficulties he has faced as an African American in his profession. He has good advice for young minority students: "Focus on your devotion to your subject. Love of your subject can give you the energy to face obstacles and challenges." Lead students to discuss Dr. Tyson's advice.

Answers to Explain It:
1. Answers will vary. Possible answer: In a city like New York, tall buildings could block a person's view of the sky. Also, the large number of bright lights would make it difficult to see many stars.

2. Answers will vary. Possible answer: In addition to exhibits that teach about the skies, a planetarium has a theater with a special projector to show movements of stars and planets and to explain certain events seen in the night sky.

3. Answers will vary. Possible answer: We tend to admire famous figures in entertainment and sports, just as we look up to the stars.

3. EXTEND

After students have completed the Explore It, ask them to share what they have learned. Have students work in small groups to find the website for the Hayden Planetarium. They can use this site or other sites to get the answers to the questions they posed.

Career Focus: Invite students to imagine that they are in charge of setting up an astronomy exhibit in a local museum, and have them draw a picture of the exhibit. What are some important ideas that students could learn from your exhibit?

J. Ernest Wilkins, Jr.

AT A GLANCE

born 1923
Chicago, Illinois

education
B.S., M.S., Ph.D., U. of Chicago
M.M.E., New York University

occupation
Physicist, Mathematician, Engineer

At the age of 13, Ernest Wilkins entered college. He graduated at 17 and earned a Ph.D. in mathematics at 19.

After graduating, though, Dr. Wilkins faced a setback. No research university would hire an African American. Instead, he took a job as a teacher at Tuskegee Institute.

Dr. Wilkins soon turned his talents to nuclear power. With a team of scientists at the University of Chicago, he investigated Plutonium 239, a material used to make the atomic bomb. When that project was completed, Dr. Wilkins worked as a mathematician for different industries. He then became an owner of a company that developed nuclear reactors to generate electric power.

Protective shields for radioactive materials

Working with nuclear power has some risks. Gamma rays, which can be harmful to humans, are released by the sun. They are also released during the process that produces nuclear power. Dr. Wilkins wanted to find materials that could shield workers from gamma rays. He developed a way to calculate the amount of radiation that different materials can absorb. His technique is now used to protect people who research nuclear materials.

? explain it

1. How can you tell that J. Ernest Wilkins was an eager student?
2. Why do researchers in space need to shield themselves from the sun?
3. If you could talk to Dr. Wilkins, what would you ask him about his work?

! explore it

How do nuclear power plants generate electricity? Use an encyclopedia or other reference book to find the answer.

J. Ernest Wilkins, Jr.
TEACHING NOTES

Purpose: To introduce students to a talented physicist, mathematician, and nuclear engineer.

Science Background: Gamma rays are high-energy rays given off from the nuclei of certain types of atoms. They are emitted when radioactive elements disintegrate. Small amounts of gamma rays are absorbed into our bodies naturally from the air we breathe and the water we drink. Large amounts of gamma rays, however, can be dangerous and damage living tissues.

People who work with radiation must protect themselves against dangerous levels of exposure. Maintaining as much distance as possible from the source of radiation is the best protection. Limiting the time a person is exposed is also helpful. However, people working in space are above the protection of Earth's atmosphere. In these cases, lead shields, which will absorb the rays, are necessary. A one-half-inch shield of lead will stop gamma rays.

Used in controlled situations, gamma rays provide benefits in many areas. Gamma rays can be used in industry to inspect metal. They have some uses in medical procedures. They also are used in preparing some food products.

1. INTRODUCE

About J. Ernest Wilkins, Jr.: Ernest Wilkins is considered a genius. He started college at the age of 13, becoming the youngest student ever to attend the University of Chicago. He received his degree in mathematics when he was 17. Dr. Wilkins was named to the National Academy of Engineering. He was the second African American to be included.

About Wilkins' Work: Dr. Wilkins is talented in many significant areas of research. His work on the Manhattan Project has led him to an enduring interest in exploring peacetime uses of nuclear energy. Dr. Wilkins served as a mathematician for the American Optical Company and, later, for the

Nuclear Development Corporation of America. He is, perhaps, best known for a mathematical model that calculates the level of radiation absorption by different materials, a technique still used to make shielding for research workers in space and on nuclear projects. Dr. Wilkins has received many awards, including the medal given to him in 1980 by the U.S. Army for Outstanding Civilian Service.

2. DISCUSS

Use the copy master to prompt discussion. Lead students to discuss how science and mathematics often work together in laboratories and in everyday life. Ask students if recent school science projects required them to take measurements, make calculations, or use math in any other way.

Answers to Explain It:
1. Answers will vary. Possible answer: Only a dedicated and hard-working student could enter a university at the age of 13 and maintain the enthusiasm and level of work necessary to earn a Ph.D. by 19.

2. Answers will vary. Possible answer: Because they are nearer the sun than on Earth, they need special protection against gamma rays.

3. Answers will vary. Possible answer: Students might want to inquire about Dr. Wilkins' work in different research laboratories.

3. EXTEND

After students have completed the Explore It, ask them to share what they have found. Students may enjoy drawing cartoon strips to illustrate what they have learned about the process by which nuclear reactors produce electrical energy.

Career Focus: Dr. Wilkins' career has included many different types of jobs, such as teacher, mathematics researcher, and manager of a research department. What skills and attitudes would a person need to be successful in all of those occupations?